Angela Pattatucci Aragón, PhD
Editor

Challenging Lesbian Norms: Intersex, Transgender, Intersectional, and Queer Perspectives

Challenging Lesbian Norms: Intersex, Transgender, Intersectional, and Queer Perspectives has been co-published simultaneously as *Journal of Lesbian Studies*, Volume 10, Numbers 1/2 2006.

Pre-publication
REVIEWS,
COMMENTARIES,
EVALUATIONS . . .

"AN INVALUABLE RESOURCE– keenly argued and passionately felt. Some readers will hate this book, some will love it, but few will find themselves able to stop thinking about it long after they put it down."

Riki Wilchins
Author of Read my Lips *and* Queer Theory/Gender Theory
Co-editor of GenderQueer

Harrington Park Press

Challenging Lesbian Norms: Intersex, Transgender, Intersectional, and Queer Perspectives

Challenging Lesbian Norms: Intersex, Transgender, Intersectional, and Queer Perspectives has been co-published simultaneously as *Journal of Lesbian Studies*, Volume 10, Numbers 1/2 2006.

Challenging Lesbian Norms: Intersex, Transgender, Intersectional, and Queer Perspectives, edited by Angela Pattatucci Aragón, PhD (Vol. 10, No. 1/2, 2006). *"An invaluable resource–keenly argued and passionately felt. Some readers will hate this book, some will love it, but few will find themselves able to stop thinking about it long after they put it down."* (Riki Wilchins, author of Read my Lips *and* Queer Theory/Gender Theory; *co-editor of* GenderQueer)

Lesbian Academic Couples, edited by Michelle Gibson and Deborah T. Meem (Vol. 9, No. 4, 2005). *"The writers gathered here expose the underlying currents that allow them to flourish and continue to grow—opportunity, activism, and great love: for their work, for justice, and for each other."* (Chris Cuomo, PhD, Professor of Philosophy and Women's Studies, University of Cincinnati)

Making Lesbians Visible in the Substance Use Field, edited by Elizabeth Ettorre (Vol. 9, No. 3, 2005). *"This is the book that we in the substance abuse treatment and research fields have been waiting for."* (Katherine van Wormer, PhD, MSSW, Professor of Social Work, University of Iowa; co-author, Addiction Treatment: A Strengths Perspective)

Lesbian Communities: Festivals, RVs, and the Internet, edited by Esther Rothblum and Penny Sablove (Vol. 9, No. 1/2, 2005). *"Important. . . . Challenging and compelling. . . . A fascinating assortment of diverse perspectives on just what defines a lesbian 'community,' what needs and desires they meet, and how those worlds intersect with other groups and cultures."* (Diane Anderson-Minshall, Executive Editor, Curve *Magazine*)

Lesbian Ex-Lovers: The Really Long-Term Relationships, edited by Jacqueline S. Weinstock and Esther D. Rothblum (Vol. 8, No. 3/4, 2004). *"Compelling. . . . In these heady days of legal gay marriage, this book is a good reminder of the devotion lesbians have always had to the women we've loved, and the vows we've made with our hearts, long before we demanded licenses. This book is a tribute to the long memory we have of the women's hands who have touched our most vulnerable parts, and the invisible hands that outlast our divorces."* (Arlene Istar Lev, CSW-R, CSAC, Author of Transgender Emergence *and* The Complete Lesbian and Gay Parenting Guide; *Founder and Clinical Director, Choices Counseling and Consulting*)

Lesbians, Feminism, and Psychoanalysis: The Second Wave, edited by Judith M. Glassgold and Suzanne Iasenza (Vol. 8, No. 1/2, 2004). *"This book is the first to set the tone for a lesbian psychoanalytic revolution."* (Dany Nobus, PhD, Senior Lecturer in Psychology and Psychoanalytic Studies, Brunel University, United Kingdom)

Trauma, Stress, and Resilence Among Sexual Minority Women: Rising Like the Phoenix, edited by Kimberly F. Balsam, PhD (Vol. 7, No. 4, 2003). *Provides a first-hand look at the victimization experiences that lesbian and bisexual women face as well as how they work through these challenges and emerge resilient.*

Latina Lesbian Writers and Artists, edited by María Dolores Costa, PhD (Vol. 7, No. 3, 2003). *"A fascinating journey through the Latina lesbian experience. It brings us stories of exile, assimilation, and conflict of cultures. The book takes us to the Midwest, New York, Chicana Borderlands, Mexico, Argentina, and Spain. It succeeds at showing the diversity within the Latina lesbian experience through deeply feminist testimonials of life and struggle."* (Susana Cook, performance artist and playwright)

Lesbian Rites: Symbolic Acts and the Power of Community, edited by Ramona Faith Oswald, PhD (Vol. 7, No. 2, 2003). *"Informative, enlightening, and well written . . . illuminates the range of lesbian ritual behavior in a creative and thorough manner. Ramona Faith Oswald and the contributors to this book have done scholars and students of ritual studies an important service by demonstrating the power, pervasiveness, and performative nature of lesbian ritual practices."* (Cele Otnes, PhD, Associate Professor, Department of Business Administration, University of Illinois)

Mental Health Issues for Sexual Minority Women: Redefining Women's Mental Health, edited by Tonda L. Hughes, RN, PhD, FAAN, Carrol Smith, RN, MS, and Alice Dan, PhD (Vol. 7, No. 1, 2003). *A rare look at mental health issues for lesbians and other sexual minority women.*

Addressing Homophobia and Heterosexism on College Campuses, edited by Elizabeth P. Cramer, PhD (Vol. 6, No. 3/4, 2002). *A practical guide to creating LGBT-supportive environments on college campuses.*

Femme/Butch: New Considerations of the Way We Want to Go, edited by Michelle Gibson and Deborah T. Meem (Vol. 6, No. 2, 2002). *"Disrupts the fictions of heterosexual norms. . . . A much-needed examiniation of the ways that butch/femme identitites subvert both heteronormativity and 'expected' lesbian behavior." (Patti Capel Swartz, PhD, Assistant Professor of English, Kent State University)*

Lesbian Love and Relationships, edited by Suzanna M. Rose, PhD (Vol. 6, No. 1, 2002). *"Suzanna Rose's collection of 13 essays is well suited to prompting serious contemplation and discussion about lesbian lives and how they are–or are not–different from others. . . . Interesting and useful for debunking some myths, confirming others, and reaching out into new territories that were previously unexplored." (Lisa Keen, BA, MFA, Senior Political Correspondent, Washington Blade)*

Everyday Mutinies: Funding Lesbian Activism, edited by Nanette K. Gartrell, MD, and Esther D. Rothblum, PhD (Vol. 5, No. 3, 2001). *"Any lesbian who fears she'll never find the money, time, or support for her work can take heart from the resourcefulness and dogged determination of the contributors to this book. Not only do these inspiring stories provide practical tips on making dreams come true, they offer an informal history of lesbian political activism since World War II." (Jane Futcher, MA, Reporter,* Marin Independent Journal, *and author of* Crush, Dream Lover, *and* Promise Not to Tell)

Lesbian Studies in Aotearoa/New Zealand, edited by Alison J. Laurie (Vol. 5, No. 1/2, 2001). *These fascinating studies analyze topics ranging from the gender transgressions of women passing as men in order to work and marry as they wished to the effects of coming out on modern women's health.*

Lesbian Self-Writing: The Embodiment of Experience, edited by Lynda Hall, PhD (Vol. 4, No. 4, 2000). *"Probes the intersection of love for words and love for women. . . . Luminous, erotic, evocative." (Beverly Burch, PhD, psychotherapist and author,* Other Women: Lesbian/Bisexual Experience and Psychoanalytic Views of Women *and* On Intimate Terms: The Psychology of Difference in Lesbian Relationships)

'Romancing the Margins'? Lesbian Writing in the 1990s, edited by Gabriele Griffin, PhD (Vol. 4, No. 2, 2000). *Explores lesbian issues through the mediums of books, movies, and poetry and offers readers critical essays that examine current lesbian writing and discuss how recent movements have tried to remove racist and antigay themes from literature and movies.*

From Nowhere to Everywhere: Lesbian Geographies, edited by Gill Valentine, PhD (Vol. 4, No. 1, 2000). *"A significant and worthy contribution to the ever growing literature on sexuality and space. . . . A politically significant volume representing the first major collection on lesbian geographies. . . . I will make extensive use of this book in my courses on social and cultural geography and sexuality and space." (Jon Binnie, PhD, Lecturer in Human Geography, Liverpool, John Moores University, United Kingdom)*

Lesbians, Levis and Lipstick: The Meaning of Beauty in Our Lives, edited by Jeanine C. Cogan, PhD, and Joanie M. Erickson (Vol. 3, No. 4, 1999). *Explores lesbian beauty norms and the effects these norms have on lesbian women.*

Lesbian Sex Scandals: Sexual Practices, Identities, and Politics, edited by Dawn Atkins, MA (Vol. 3, No. 3, 1999). *"Grounded in material practices, this collection explores confrontation and coincidence among identity politics, 'scandalous' sexual practices, and queer theory and feminism. . . . It expands notions of lesbian identification and lesbian community." (Maria Pramaggiore, PhD, Assistant Professor, Film Studies, North Carolina State University, Raleigh)*

The Lesbian Polyamory Reader: Open Relationships, Non-Monogamy, and Casual Sex, edited by Marcia Munson and Judith P. Stelboum, PhD (Vol. 3, No. 1/2, 1999). *"Offers reasonable, logical, and persuasive explanations for a style of life I had not seriously considered before. . . . A terrific read." (Beverly Todd, Acquisitions Librarian, Estes Park Public Library, Estes Park, Colorado)*

Challenging Lesbian Norms:
Intersex, Transgender, Intersectional, and Queer Perspectives

Angela Pattatucci Aragón
Editor

Challenging Lesbian Norms: Intersex, Transgender, Intersectional, and Queer Perspectives has been co-published simultaneously as *Journal of Lesbian Studies*, Volume 10, Numbers 1/2 2006.

HPP

Harrington Park Press®
An Imprint of The Haworth Press, Inc.

New York • London • Victoria (AU)
www.HaworthPress.com

Published by

Harrington Park Press®, 10 Alice Street, Binghamton, NY 13904-1580 USA

Harrington Park Press® is an imprint of The Haworth Press, Inc., 10 Alice Street, Binghamton, NY 13904-1580 USA.

Challenging Lesbian Norms: Intersex, Transgender, Intersectional, and Queer Perspectives has been co-published simultaneously as *Journal of Lesbian Studies*, Volume 10, Numbers 1/2 2006.

The development, preparation, and publication of this work has been undertaken with great care. However, the publisher, employees, editors, and agents of The Haworth Press and all imprints of The Haworth Press, Inc., including The Haworth Medical Press® and The Pharmaceutical Products Press®, are not responsible for any errors contained herein or for consequences that may ensue from use of materials or information contained in this work. With regard to case studies, identities and circumstances of individuals discussed herein have been changed to protect confidentiality. Any resemblance to actual persons, living or dead, is entirely coincidental.

The Haworth Press is committed to the dissemination of ideas and information according to the highest standards of intellectual freedom and the free exchange of ideas. Statements made and opinions expressed in this publication do not necessarily reflect the views of the Publisher, Directors, management, or staff of The Haworth Press, Inc., or an endorsement by them.

Cover design by Marylouise E. Doyle

Library of Congress Cataloging-in-Publication Data

Challenging lesbian norms : intersex, transgender, intersectional, and queer perspectives / Angela Pattatucci Aragon, editor.
     p. cm.
     "Co-published simultaneously as Journal of lesbian studies, volume 10, numbers 1/2 2006."
     Includes bibliographical references and index.
     ISBN-13: 978-1-56023-644-3 (hard cover : alk. paper)
     ISBN-10: 1-56023-644-2 (hard cover : alk. paper)
     ISBN-13: 978-1-56023-645-0 (soft cover : alk. paper)
     ISBN-10: 1-56023-645-0 (soft cover : alk. paper)
     1. Lesbianism. 2. Transsexualism. 3. Hermaphroditism 4. Lesbians–Social conditions.
5. Lesbians–Psychology. I. Aragon, Angela Pattatucci. II. Journal of lesbian studies.
HQ75.5.C429 2006
306.76'63–dc22
                                                   2005033374

Indexing, Abstracting & Website/Internet Coverage

This section provides you with a list of major indexing & abstracting services and other tools for bibliographic access. That is to say, each service began covering this periodical during the year noted in the right column. Most Websites which are listed below have indicated that they will either post, disseminate, compile, archive, cite or alert their own Website users with research-based content from this work. (This list is as current as the copyright date of this publication.)

Abstracting, Website/Indexing Coverage . . . . Year When Coverage Began

- *Abstracts in Social Gerontology: Current Literature on Aging* **1997**

- *British Library Inside (The British Library) <http://www.bl.uk/services/current/inside.html>* **2006**

- *Cambridge Scientific Abstracts is a leading publisher of scientific information in print journals, online databases, CD-ROM and via the Internet <http://www.csa.com>* **2006**

- *Contemporary Women's Issues* **1998**

- *EBSCOhost Electronic Journals Service (EJS) <http://ejournals.ebsco.com>* **2001**

- *Elsevier Scopus <http://www.info.scopus.com>* **2005**

- *Family & Society Studies Worldwide <http://www.nisc.com>* **2001**

- *Family Index Database <http://www.familyscholar.com>* **2003**

- *Feminist Periodicals: A Current Listing of Contents* **1997**

- *GenderWatch <http://www.slinfo.com>* **1999**

- *GLBT Life, EBSCO Publishing <http://www.epnet.com/academic/glbt.asp>* **2004**

(continued)

- *Google <http://www.google.com>* **2004**
- *Google Scholar <http://www.scholar.google.com>* **2004**
- *Haworth Document Delivery Center*
  <http://www.HaworthPress.com/journals/dds.asp> **1997**
- *HOMODOK/"Relevant" Bibliographic database,*
  Documentation Centre for Gay & Lesbian Studies,
  University of Amsterdam (selective printed abstracts
  in "Homologie" and bibliographic computer databases
  covering cultural, historical, social & political aspects)
  <http://www.ihlia.nl/>.................................... **1997**
- *(IBR) International Bibliography of Book Reviews on the Humanities*
  and Social Sciences (Thomson) <http://www.saur.de> **2006**
- *IBZ International Bibliography of Periodical Literature*
  on the Humanities and Social Sciences (Thomson)
  <http://www.saur.de>.................................... **2001**
- *IGLSS Abstracts <http://www.iglss.org>*...................... **2000**
- *Index to Periodical Articles Related to Law*
  <http://www.law.utexas.edu>............................ **1997**
- *Internationale Bibliographie der geistes- und*
  sozialwissenschaftlichen Zeitschriftenliteratur . . . See IBZ
  <http://www.saur.de>.................................... **2001**
- *Lesbian Information Service*
  <http://www.lesbianinformationservice.org> **2003**
- *Links@Ovid (via CrossRef targeted DOI links)*
  <http://www.ovid.com>.................................... **2005**
- *Magazines for Libraries (Katz) (Bibliographic Access)*
  . . . (see 2003 edition).................................... **2003**
- *MEDLINE (National Library of Medicine)*
  <http://www.nlm.nih.gov> **2004**
- *Ovid Linksolver (OpenURL link resolver via CrossRef targeted*
  DOI links) <http://www.linksolver.com> **2005**
- *Psychological Abstracts (PsycINFO) <http://www.apa.org>* **2001**
- *Psychology Today* **1999**
- *Public Affairs Information Service (PAIS) International*
  (Cambridge Scientific Abstracts)
  <http://www.pais.org/www.csa.com>...................... **1997**
- *PubMed <http://www.ncbi.nlm.nih.gov/pubmed/>*.............. **2004**
- *Referativnyi Zhurnal (Abstracts Journal of the All-Russian Institute*
  of Scientific and Technical Information–in Russian)
  <http://www.viniti.ru>.................................... **1997**

(continued)

- *Scopus (see instead Elsevier Scopus)*
  <http://www.info.scopus.com> **2005**

- *Sexual Diversity Studies: Gay, Lesbian, Bisexual & Transgender*
  Abstracts (EBSCO)
  <http://www.epnet.com/acdemic/acasearchprem.asp> **2003**

- *Social Services Abstracts (Cambridge Scientific Abstracts)*
  <http://www.csa.com> **1998**

- *SocIndex (EBSCO)* **2006**

- *Sociological Abstracts (Cambridge Scientific Abstracts)*
  <http://www.csa.com> **1998**

- *Studies on Women and Gender Abstracts*
  <http://www.tandf.co.uk/swa> **1998**

- *zetoc (The British Library) <http://www.bl.uk>* **2004**

Special Bibliographic Notes related to special journal issues (separates) and indexing/abstracting:

- indexing/abstracting services in this list will also cover material in any "separate" that is co-published simultaneously with Haworth's special thematic journal issue or DocuSerial. Indexing/abstracting usually covers material at the article/chapter level.
- monographic co-editions are intended for either non-subscribers or libraries which intend to purchase a second copy for their circulating collections.
- monographic co-editions are reported to all jobbers/wholesalers/approval plans. The source journal is listed as the "series" to assist the prevention of duplicate purchasing in the same manner utilized for books-in-series.
- to facilitate user/access services all indexing/abstracting services are encouraged to utilize the co-indexing entry note indicated at the bottom of the first page of each article/chapter/contribution.
- this is intended to assist a library user of any reference tool (whether print, electronic, online, or CD-ROM) to locate the monographic version if the library has purchased this version but not a subscription to the source journal.

Challenging Lesbian Norms: Intersex, Transgender, Intersectional, and Queer Perspectives

CONTENTS

Introduction: Challenging Lesbian Normativity   1
    Angela Pattatucci Aragón

Fleshy Specificity: (Re)considering Transsexual Subjects
    in Lesbian Communities   17
    Kelly Coogan

The Invisible Body of Queer Youth: Identity and Health
    in the Margins of Lesbian and Trans Communities   43
    Dorinda L. Welle
    Sebastian S. Fuller
    Daniel Mauk
    Michael C. Clatts

"Gee, I Didn't Get That Vibe from You": Articulating My Own
    Version of a Femme Lesbian Existence   73
    Robbin VanNewkirk

Gender Please, Without the Gender Police: Rethinking Pain
    in Archetypal Narratives of Butch, Transgender,
    and FTM Masculinity   87
    Madelyn Detloff

Household Remedies: New Narratives of Queer Containment
    in the Television Movie   107
    Cait Keegan

"My Spirit in My Heart": Identity Experiences and Challenges
    Among American Indian Two-Spirit Women   125
    Karina L. Walters
    Teresa Evans-Campbell
    Jane M. Simoni
    Theresa Ronquillo
    Rupaleem Bhuyan

Teaching Transgender in Women's Studies: Snarls and Strategies 151
    Sara E. Cooper
    Connor James Trebra

Developing an Identity Model for Transgender and Intersex
    Inclusion in Lesbian Communities   181
    Christopher Robinson

An *Other* Space: Between and Beyond Lesbian-Normativity
    and Trans-Normativity   201
    Myfanwy McDonald

In Another Bracket: Trans Acceptance in Lesbian Utopia   215
    Jamie Stuart

Debating Trans Inclusion in the Feminist Movement:
    A Trans-Positive Analysis   231
    Eli R. Green

I Don't Know Who I Am: Severely Mentally Ill Latina WSW
    Navigating Differentness   249
    Sana Loue
    Nancy Méndez

Index   267

ABOUT THE EDITOR

Angela Pattatucci Aragón, PhD, is Director of the Center for Evaluation and Sociomedical Research (CIES, by its initials in Spanish) and Professor of Health Services Research Administration at the University of Puerto Rico, School of Public Health. She completed her baccalaureate studies in Education and Biology at Northeastern Illinois University in Chicago, her doctoral studies in Behavior Genetics at Indiana University–Bloomington, and her postdoctoral studies in Epidemiology and Adolescent Development at the National Institutes of Health in Bethesda, Maryland. During her postdoctoral period, she gained international recognition for her collaboration with Dean Hamer on investigating potential genetic contributions to sexual orientation. This work was featured in major newpapers, periodicals (e.g., *Time, Newsweek, U.S. News and World Report, Atlantic Monthly*), and major television networks throughout the world. Dr. Pattatucci Aragón has a long track record of experience working in areas surrounding health and policy issues focused on Latinos, women, and sexual minorities. She maintains an active interest in research on gender and sexuality, in particular work with developing adolescent sexuality and unique challenges faced by gay, lesbian, bisexual, transgender, and intersex youth. Her edited book *Women in Science: Meeting Career Challenges* has been used as a tool to guide policy in university science departments and as a learning aid in Women's Studies courses. She has authored several articles and book chapters on sexual orientation and its development. Complementing this work are strong interests in risk prevention in low-income urban youth and young adults, specifically in the areas of HIV/AIDS, substance abuse, and violence prevention. Dr. Pattatucci Aragón currently collaborates on several projects, including a study with CIES researchers investigating treatment needs for drug dependence and prevention of HIV and hepatitis B/C in Puerto Rican prisons and an epidemiological study investigating asymptomatic or unreported gonorrhea and chlamydia infections throughout the island. Other collaborative studies have demonstrated a strong link between childhood sexual abuse and adult substance abuse problems in Puerto Rican women. Dr. Pattatucci Aragón also directs a large integrated needs assessment aimed at providing information on the existing infrastructure and capacity needs in the public health system throughout Puerto Rico.

Introduction:
Challenging Lesbian Normativity

Angela Pattatucci Aragón

There is no word in our language that can describe what we are.

–*Fire*, Zeitgeist Films, 1996[1]

My journey with this volume began in March 2000, when I was asked to be a panel member at a Workshop on Lesbian Health sponsored by the U.S. Department of Health and Human Services (DHHS). A diverse group of public health professionals, community leaders, and activists had been brought together to develop strategies for implementing recommendations contained in the Institute of Medicine's report on Lesbian Health.[2]

Although I had come out as a lesbian decades earlier, I remained deeply in the closet about being intersex. Throughout the years I had found that there simply was no good way to approach the subject. Telling people early in the friendship stage tended to shut down communication entirely. Waiting for friendships to develop before I revealed my intersex status usually meant that I was accused of being disingenuous.

Angela Pattatucci Aragón is Director of the Center for Evaluation and Sociomedical Research (CIES, by its initials in Spanish) and Professor of Health Services Research Administration at the University of Puerto Rico, School of Public Health.

[Haworth co-indexing entry note]: "Introduction: Challenging Lesbian Normativity." Pattatucci Aragón, Angela. Co-published simultaneously in *Journal of Lesbian Studies* (Harrington Park Press, an imprint of The Haworth Press, Inc.) Vol. 10, No. 1/2, 2006, pp. 1-15; and: *Challenging Lesbian Norms: Intersex, Transgender, Intersectional, and Queer Perspectives* (ed: Angela Pattatucci Aragón) Harrington Park Press, an imprint of The Haworth Press, Inc., 2006, pp. 1-15. Single or multiple copies of this article are available for a fee from The Haworth Document Delivery Service [1-800-HAWORTH, 9:00 a.m. - 5:00 p.m. (EST). E-mail address: docdelivery@haworthpress.com].

The essence of my dilemma is that aside from the umbrella term *intersex*, the stigmatizing and sensationalizing term *hermaphrodite*, and medical terminology that reduces me, and others like me, to *ambiguous genitalia* or designations for genetic mutations, no words exist that can adequately describe who and what I am. Additionally, the fact that the public really has not been educated about intersex persons means that there is neither a common background nor an adequate language with which to begin a dialogue. Each conversation must therefore start with Intersex 101, a practice that focuses on *my condition* rather than on me as a person. Thus, bound by shame and self-loathing, I went through life employing a survival strategy that I had learned from adults as a teenage girl when my intersex status was revealed–to outright lie to friends and acquaintances by constructing false histories in order to protect them from the horror of *my condition*. Of course, this lack of authenticity caused me to feel even worse about myself and fed the cycle of shame and self-loathing.

I wanted to live free, to shed the yoke of intersex disgrace that I had carried for so long, but I also wanted to belong–to be part of the lesbian family. A major roadblock in my quest for freedom was the sorority-like character that is common in lesbian communities. Sororities, and analogously lesbian communities, are by definition exclusive spaces in which women typically pass through a harsh and subjective social-selection process that involves learning to align beliefs and behaviors to the majority in lieu of achieving a privileged membership status. Individuality is compromised; women reinvent themselves within specific boundaries set by the sorority or lesbian community. The result is a reification of patriarchal social hierarchies, whereby members of the sorority or lesbian community enjoy a privileged status relative to those women deemed unworthy of membership. In both lesbian communities and sororities, conformity is enforced through the threat of ostracism. This point in particular terrified me. I understood only too well what would happen if I exited my closet.

> She dumped me. . . . She wanted to be with somebody a little more . . . a little more gay.
>
> –*Kissing Jessica Stein*, 20th Century Fox, 2002

A certain sanctioned hypocrisy exists in lesbian communities with respect to gender difference. Each year at Gay Pride celebrations and at other gatherings we clamor for human rights, social justice, and respect for diversity. Yet, over the years, lesbian communities have had "an awfully bad reputation of defining what is authentically female or lesbian, restricting access to the small range of rights and privileges won by the gay and lesbian movement, and making negative judgments on nonconformity."[3] I had burned bridges to the mainland straight community years before and I perceived that banishment from the Isle of Lesbos would leave me with no other alternative but to drown in a sea of loneliness and despair. I therefore decided to tell no one about my intersex status and continue the practice of concocting lies about my history. However, maintaining my *secret* meant keeping myself at arm's length from everyone. I traveled in lesbian circles projecting what amounted to a hologram of the individual I perceived that I was expected to be, constantly assessing people's reactions and making adjustments as necessary. It was exquisite control in a context of excruciating loneliness. Consequently, I really did not have to wait for lesbian xenophobes to discover my *true nature* and ostracize me. I had, in a sense, already condemned myself to a life in exile.[4]

The underlying theme of my story–extreme anxiety over coming out in a highly normalized and value-laden community setting–will be so familiar to most that it is unnecessary for me to provide any more details. Nevertheless, my path diverges with respect to the experiences of my other lesbian sisters in that my agonizing struggles over coming out occurred within the context of *lesbian communities* rather than straight ones. Like many of you growing up hearing homophobic remarks and remaining silent out of fear, I listened to lesbians make derogatory statements and disrespectful jokes about trans[5] people at conferences and at other gatherings (intersex people still are not really part of the equation). I listened to lesbian feminists allege conspiracy theories about transwomen acting as government agents seeking to infiltrate lesbian (i.e., real women's) spaces. I heard all manner of trans character assassination, stereotyping and caricatures, and a host of accusations made about the so-called trans agenda to undermine lesbian communities.[6] With each comment, I felt my stomach contort. Yet through it all, I sat quietly and said nothing, fearful that speaking out would lead to negative labeling and being ostracized. Sound familiar?

Although the thread that linked me to lesbian communities was tenuous at best, I held onto it as if it were a lifeline. Petrified by an overwhelming fear of rejection, I allowed myself to be stuffed with vitriolic anti-trans messages, month after month, year after year, until finally, at the DHHS Workshop on Lesbian Health, I could not swallow another helping of anti-trans rancor and I vomited, both figuratively and literally.

Fortunately, other trans and intersex individuals were a lot more courageous than I was and had spoken out against lesbian exclusionary practices. A few of those individuals were in attendance at the DHHS Workshop. They routinely called attention to the fact that the presentations and discussions were overlooking health issues important to trans people within lesbian communities. Unfortunately, these requests were met with curt, passive-aggressive, and condescending responses. My anger seethed. I also was frustrated by the fact that the trans activists in attendance presented female-to-male (FTM) issues as if they were the only important trans concerns within lesbian communities. I felt erased.

Following lunch on the first day, we were scheduled to break into panels to strategize about specific areas of lesbian health concern. While waiting for the rest of the group to arrive at our session, I engaged in small talk with a few of the other panel members. Still fuming over the dismissal of trans concerns earlier in the day, I asked for opinions about issues raised by the trans activists. The responses I received were not positive. Trans activists at the workshop were portrayed as either selfishly diluting genuine lesbian health concerns or trying to hijack the entire workshop. Undaunted, I raised the stakes and introduced the invisible intersex issue in lesbian communities. The response in this case was a brush off.

"We can't be expected to take on the needs of every splinter group," one of the women responded. "Let them organize their own workshop. This one is for us."

I politely reminded the women that there are intersex and trans persons *within* lesbian communities and that these issues also should be lesbian concerns. However, by this time some in the group were losing patience and had become defensive. A woman who previously had been silent snapped at me with what could be best described as a paraphrase of the military's don't ask, don't tell policy. She informed me that intersex and trans people wishing to be part of lesbian communities have a responsibility to table their issues in lieu of advancing a lesbian-specific agenda. At this point, I lost it. I had become so disillusioned by the near complete obliviousness of these women to their privileged status that I came out as intersex then and there. The reactions ranged from shock to sneers.

Fueled by indignation, I did not stop with coming out. I criticized the ludicrousness of spending so much energy on defining what is authentically female or lesbian, using my own life pedagogically as an ever-changing and expanding object of knowledge that instructs. I was assigned female at birth and raised as such, thus meeting the infamous *womyn-born womyn* or *lesbian born female* criterion applied in certain lesbian circles throughout the world. However, I began undergoing a *naturally occurring* virilization process[7] at puberty that, had it not been stopped, would have resulted in a near complete transformation from female to male.

"We often speak metaphorically of testosterone poisoning," I said. "Well, I'm a living example of it. Now that you know, does this mean that in your eyes I am suddenly no longer a real female, a real lesbian?"

The fact that my virilization process was halted by doctors meant that I had to undergo (more accurately endure) psychiatric evaluation with accompanying surgical intervention to *reconfirm* my original sex assignment as female at birth. This process contained elements similar to those applied to male-to-female (MTF) transsexuals. Thus, although none of it was by choice, I have traversed a gamut from womyn-born womyn, to a brief period analogous to FTM transitioning, to *corrective* interventions that mirror those experienced by MTF transsexuals. Mapped onto this trajectory has been a long-standing lesbian sexual orientation.

As I continued my discourse, the group surrounding me grew. I was a highly visible federal government official at the time, directing scientific review of HIV/AIDS grant applications for the National Institutes of Health, and I think that people were astonished that I was willing to come out about my intersex status at a gathering where there were other federal officials in attendance. Looking back, I am kind of astonished myself. I did it because, most of all, I wanted to be understood and I was sick and tired of living in the closet and being surrounded by lesbian hypocrisy. I talked to the group about how dehumanizing it is to sit at gatherings and listen to women dismiss entire classes of people as irrelevant because we do not fit nicely into their mainstream lesbian political agenda. Equally dehumanizing is having sit and listen to women refer to trans issues only as an afterthought, something that is tacked onto the end of sentences out of political correctness rather than genuine concern, and to not even hear the word intersex spoken in discussions.

When I finished, several in the group thanked me for sharing and some even complimented my courage. I then sat down emotionally drained, but ready to contribute to the important work of the panel. A gay male HIV/AIDS researcher seated next to me leaned over and whispered into my ear.

"I think you've just committed suicide."

"Oh, I don't think there will be any repercussions on my job," I replied.

"No. I mean you committed suicide with them," he said, gesturing to the women in the room attending the session.

It turns out that he was right. I was given the silent treatment for the remainder of the two-day workshop. Women (also the FTM trans activists) suddenly were unavailable to share meals or to talk during breaks. I was told that empty seats at the workshop were being saved for other participants that I noticed remained empty after I had moved on. In short, I was banished to spend my time with the few gay males in attendance or with other presumed heterosexual federal officials.

Esther Rothblum, editor of the *Journal of Lesbian Studies*, was in attendance at the workshop. A friend and colleague, she did not participate in the ostracizing behavior of other the attendees and instead encouraged me to consider developing a special *JLS* issue on intersex, transgender, intersectional, and queer issues in lesbian communities. However, my status as a federal official meant that any material that I produced had to pass through multiple levels of federal oversight before it finally could be submitted for publication.[8] Assuming that the proposed special *JLS* issue never would pass through federal scrutiny, I politely declined.

I wish that I could say that my experience at the DHHS workshop was a catalyst for redefining my life's direction toward intersex and trans activism. It was not. Instead, I quietly slipped back into my closet and bolted the door when the workshop ended. I remained there for another year before coming out once and for all.

Being only one among just a few token lesbians at the NIH, I had been assigned to give a workshop on strategies for preparing successful federal grant applications at the National Lesbian Health Conference in San Francisco. Cheryl Chase, Founder and Executive Director of the Intersex Society of North America, was in attendance and approached me after my presentation. Her mere presence was like a spotlight exposing my closeted life. I felt ashamed and came out to her right away. As they say, once the toothpaste is squeezed out of the tube, you can't push it back in. I continued coming out to people and I got involved. Cheryl has that kind of affect on people.

A few months later, I resigned my position with the federal government and returned home to Puerto Rico. This opened up the possibility to revisit Esther Rothblum's original invitation to develop a special *JLS* issue on intersex, transgender, intersectional, and queer issues in lesbian communities. With her support, I set out to recruit the authors that have contributed articles to this volume. I hope that you find them engaging and valuable.

> You're right. We are different. But not in the way that you mean. We're different because you are all ashamed of us, but we're not ashamed of you.
>
> –*Flawless*, Metro-Goldwyn-Mayer Pictures, 1999

In previous decades, the heated debate over trans inclusion in lesbian communities has been restricted to a small group of lesbian feminists and trans activists. The arguments primarily have centered on the issue of who has legitimacy in lesbian space. As I mentioned previously, I have observed many of these debates over the years and a unifying thread linking them is the use of dehumanizing language. For example, a common anti-trans statement that was circulating around lesbian communities a few years ago was that you can use a hypodermic needle to squirt apple juice into an orange, but that this will not change the orange into an apple. This prejudicial statement is designed to bolster assertions that trans individuals have no legitimate right to enter lesbian space because they are judged to be counterfeit women.

Ironically, lesbians have been on the disenfranchised end of a legitimacy debate with the mainstream heterosexual community for decades. Even I have written about it in the past.

> I have arrived at the conclusion through my life experiences that Lesbians are viewed by the world as lacking authenticity. We are seen as *incomplete* women, much like an unfinished painting on a canvas by a deceased artist, stuck in an inescapable limbo that becomes its existence, its reality, and everything interesting about the work–the levels of meaning in its depiction, its tone, hue, texture, lines, depth, and contour–are all relegated insignificant. Nothing matters except that the painting, or the woman in our case, is an *unfinished work*. People always wonder what we could have been, instead of celebrating what we are.[9]

Lesbians have worked tirelessly toward achieving legitimacy as human beings, as women, as mothers, and as life partners in the eyes of an obstinate heteronormative[10] majority. Lesbians have felt the pain of being judged as less than human and being deemed unworthy of basic human rights. Lesbians know the sting of discrimination in key areas such as employment, housing, and healthcare. Lesbians have shed tears at vigils for hospitalized or deceased sisters that were victims of violence and hate crimes. Given this history, the vitriol over trans inclusion in lesbian communities is difficult to understand.

While the trans inclusion issue primarily has remained an internal matter within selected lesbian communities and spaces, recent progress in areas such as same-sex marriage has caused the issue of group definition for lesbians to take on greater importance. One way to think about this is that lesbian couples are asking the heteronormative majority for permission to move up in the patriarchal hierarchy–to be considered equal to heterosexuals, to be extended the time-honored privilege of marriage that previously has been the exclusive domain of heterosexuals. With the stakes so high, it has become necessary to educate the mainstream populace regarding who we are talking about (and who we are not) when the term lesbian is used. In order to appear worthy of the marital privilege, lesbians have had to take on the daunting challenge of proving that we are not just equal to the heterosexual norm, but that we are better–better mothers, better partners, better and more stable relationships, and better families with well-adjusted children that have stellar academic records. Advancing this *better than* image has required a simultaneous effort to sanitize lesbian communities of anything deemed stereotypical and to present images that are compatible with the *canon of family values* (e.g., loving lesbian mothers cradling infants, lesbian soccer moms, lesbian couples celebrating long-term stable relationships). In the process, we are behaving much like the factory worker who sheds her dirty work clothes and dresses in a nice suit for a meeting with her boss to request consideration for a promotion. However, in this case a lot more is at stake. One could argue that the subtext of the marriage debate is that lesbians and gay men are asking the heteronormative majority to consider them fully human.[11] The problem, as I see it, is that the fight for marriage rights has led to the *dehumanization* of individuals that do not or cannot adhere to the canon of family values, only this time the primary source of marginalization is from lesbian communities rather than straight ones. The process also leads to scapegoating. Failures in the fight for same-sex marriage rights are blamed on the obstinacy of gender-variant individuals in remaining visible and insisting that their voices be heard.

The zealous focus on same-sex marriage with its accompanying sanitization efforts has placed the power to define lesbian communities largely in the hands of the heteronormative majority. The result has been an emerging *homonormativity* that combines the essentialism of the trans inclusion controversy with the desire to appear acceptable in the eyes of the heteronormative majority (e.g., we're just like you except for one thing).

I define homonormativity in lesbian communities as consisting of the following elements.

- An unwavering belief in the anatomical facticity of the female body.[12]
- A rejection of the patriarchally defined feminine gender role (with the exception of motherhood) and standards of appearance, as long as that rejection does not manifest itself in the embracing of masculine identities.[13]
- Exclusive desire and attraction to individuals meeting the first two criteria.

Intersex, transgender, intersectional, and queer individuals are viewed incompatible with this new homonormative image and as such are pushed further out on the margins of lesbian communities or are ostracized completely.

> You know why the man stuff seemed so real? Because I'm pretending. You see a man through the mirror of a woman through the mirror of a man. You take one of those reflecting glasses away; it doesn't work. The man only works because you see him in contrast to the woman he is. If you saw him without the her he lives inside, he wouldn't seem a man at all.
>
> *–Stage Beauty*, Lions Gate Films, 2004

Strong parallels exist between the same-sex marriage and the trans inclusion in lesbian communities debates. Both center on legitimacy and exclusivity. In the case of same-sex marriage, the heteronormative majority claims the institution of marriage as their exclusive space. Because they hold the power, the only legitimate marriages are defined as those occurring between opposite-sex couples.[14] Similarly, in lesbian communities certain venues and events are claimed as exclusive space for lesbians. Legitimacy to enter into lesbian-exclusive space is determined most often by the arbitrary *womyn-born womyn* criterion.

I always have found the concept of womyn-born womyn confusing. The implication is that those women failing to meet this criterion are counterfeit or inauthentic. While on the surface womyn-born womyn appears to be an essentialist concept, I am not convinced that it qualifies as such. Essentialists seek to distill the essence of things, universalize them (usually in a biological or cultural sense), and to define them so that unambiguous distinctions can be made. However, the anti-trans lesbian feminists that apply this term actually do no such thing. Instead, the term womyn-born womyn just hangs. The phrase may sound cute, but as a defining criterion it is woefully inadequate. We have no information on what aspects of the global construct *woman* are considered to be important and what aspects are deemed irrelevant when it comes to determining who is allowed to enter into lesbian space. Like the equally inadequate heteronormative criterion for marriage (a union between one man and one woman), the womyn-born womyn concept seems–dare I say it?–patriarchal. The policy is applied arbitrarily to exclude by those in power and is not subject to question or explanation.

Whether the controversy is over who has a legitimate right to marry or which persons have a legitimate right to claim a lesbian identity and enter lesbian space, the oppressive behavior is a reflection of group-based social hierarchies to exercise social dominance.[15] One group with power lords it over another disenfranchised group–the heteronormative majority over lesbians; lesbians over trans persons. In the process, fairness never enters into the equation. The only factor deemed worthy of consideration is the comfort of those in power. Thus, like its sister heteronormitivity, "homonormative hegemony is about silent access to increased comfortability–as well as *silencing* sustained efforts to critique this comfortability–on the part of homonormative lesbians."[16]

> Ah, let's see. This is the part where you tell me that what matters is on the inside. And inside of you is a little dyke just like me.
>
> –*Bound*, Republic Pictures, 1996

An informative analogy to the trans inclusion in lesbian communities issue can be found in adult support groups for individuals that were abused as children. In the absence of an effective facilitator, group members will tend to hierarchically rank each other based upon the perceived severity of abuse experienced by the various members. The result is destructive and based upon a disingenuous and politically doomed impulse to quantify victimhood. Instead of focusing on healing, the hierarchical ranking process concentrates members' energies on reliving specific details of abuse in order to justify their pain and right to be part of the group. Focusing on details establishes a competitive environment and the hierarchies perpetuate suffering. The group stagnates and the emotional and mental-health state of members declines. Conversely, focusing on healing creates a cooperative environment of mutual support. Members acknowledge that healing is the common challenge that everyone faces and a value system emerges that is based upon the tenet that no group member should have to suffer.

Lesbian, gay, bisexual, transsexual, transgender, intersex, queer, and intersectional individuals (and a host of others not mentioned) all live under the oppression of the heteronormative majority. Right now, we are behaving like the support group above that has an ineffective facilitator. We are focused on details, creating a competitive environment based upon hierarchical rankings. In short, we are oppressed and we are behaving like oppressed people. But it doesn't have to be this way. Like the example above, we can acknowledge and respect our differences while at the same time recognizing that oppression by the heteronormative majority is the common challenge that we all face. We can *choose* not to behave like oppressed people and in doing so decide that none of us should have to suffer. Healing could emerge as a deeply political locale for growth and discovery in lesbian communities. Healing could serve as the axis from which a new system of lesbian ethics might develop. Healing might act as a political catalyst for lesbian, gay, bisexual, transsexual, transgender, intersex, queer, and intersectional individuals to learn from each other and to develop new alliances.

> You wanna know what's crazy? After all these years I'm still sitting here trying to justify my life. This is crazy.
>
> –*Torch Song Trilogy*, New Line Productions, 1988

This volume challenges lesbian normativity from intersex, transgender, intersectional, and queer perspectives. In some cases, such as this introduction, the perspective is informed mostly by personal experience, in others it is by theory, and in still others it is by field research. In developing the volume, I purposely sought contributions that explored homonormativity and its effects within lesbian communities from diverse angles. I also made a specific commitment to feature new voices and challenged contributors to approach their articles from fresh, previously unexplored directions.

I am indebted to Esther Rothblum for her enthusiastic support and patience in bringing this project to fruition. I also am grateful to Colleen Coughlin and Elizabeth Dale for their contributions and thoughtful comments in the early stages of the project. Finally, I am extremely thankful for the dedication and passion that the contributing authors brought to this volume. It has been a fantastic journey working together; a collaborative effort all the way. I am grateful for your support and friendship.

NOTES

1. I quote popular films in this paper and acknowledge that they are taken out of context. My aim in using them is not to create a direct linkage to issues raised in the text through cinematic examples. Instead, I use the quotes as metaphors both to anchor and to drive the discussion forward.

2. Institute of Medicine (1999). *Lesbian Health: Current Assessment and Directions for the Future*. Washington, D.C.: National Academies of Science. The report can be found and purchased at http://www.iom.edu/report.asp?id=5606.

3. This quote appears in Sara Cooper and Connor Trebra's article featured in this volume: "Teaching Transgender in Women's Studies: Snarls and Strategies."

4. I acknowledge my own failing in this. By choosing secrecy over disclosure with respect to my intersex status, I did not give women in lesbian communities a chance to be accepting and supportive. However, it is important to recognize that my attitudes were shaped surrounding the issue of disclosure during the late 1960s and throughout the 1970s, a time of extreme dogma-centered policing within lesbian communities. Indoctrination into lesbian communities involved what amounted to an act of contrition–purging all patriarchal influences and underpinnings in one's life and embracing *the way, the truth, and the life* of lesbian ethics. Fanatical rules were developed for virtually every aspect of life, including prohibitions against certain positions and practices used during lovemaking, wearing traditional feminine attire (including grooming practices such as shaving and hygienic measures such as the use of tampons), regulations governing modes of speech and the people with whom you associated, and a host of other things deemed patriarchal. (For a more detailed analysis of lesbian policing see Faderman L (1991). *Odd Girls and Twilight Lovers*. New York: Columbia University Press.) It was a time when women (including lesbians) with children were scorned as *breeders* (my how things have changed with respect to childbearing). Coming out within this fanatical community context was neither a power-neutral nor a value-neutral process. It was not akin to me announcing that I liked pink lipstick in a group of women all wearing red. There were real consequences. Of course, one could say that this is true for the coming out experiences of most lesbians and gay men. However, disclosure of sexual orientation for lesbians and gay men most typically involves coming out of one support system (e.g., mainstream straight community) and coming into another (e.g., underground homosexual community). This holds true for my coming out experience as a lesbian, but not as intersex. Trans, intersex, and queer movements did

not emerge until the 1990s. The other half of the equation did not exist for me; there was no other support system to *come into*. At the time, I could not see any other option except to remain in my closeted, shame-bound life.

5. I am using Eli Green's definition of *trans* as denoting a person whose gender identity is not congruent with their assigned biological sex. Eli prefers the term *trans* to transgender or transsexual because it does not inherently assume surgical or hormonal status/desire. Using *trans* also represents a purposeful move to include people whose identities do not necessarily fit within the constraints of the more commonly used terms transgender or transsexual (or who do not feel comfortable with these particular labels). See Eli Green's article featured in this volume: "Debating Trans Inclusion in the Feminist Movement: A Trans-Positive Analysis."

6. The accusation that there is a trans agenda to undermine lesbian communities has strong parallels to the irrational fears expressed by Christian conservatives alleging a homosexual agenda with anarchistic aims to destabilize the principles and values under which the United States was founded. Some conservatives go as far as to claim that the so-called homosexual agenda is bent upon destroying the world. While lesbian communities make no such broad-based claim, the rhetoric used by anti-trans lesbians is thematically similar to that of Christian conservatives, and one could argue that anti-trans lesbians *are* claiming that the so-called trans agenda would destroy *their world* as they know it.

7. Intersex females with *5-alpha reductase deficiency* or partial *androgen insensitivity syndrome* are born with a vulva and vaginal opening (thus the female sex assignment) but also with testis located in the lower abdomen where the ovaries usually would be found. If undetected, a *naturally occurring* virilization process commences at puberty when the abdominally located testis begin producing varying amounts of testosterone. If left to proceed, the testis eventually will migrate and fill the outer labial folds, the clitoris will enlarge to resemble a penis, and masculine external characteristics will develop.

8. The necessity for federal oversight was in part related to my position at the NIH and was reasonable. Because I presided over HIV/AIDS scientific review sessions, all real or perceived sources of conflict of interest had to be avoided. Since HIV/AIDS is such a broad and diverse field of research, this in effect eliminated all possibilities for academic collaboration. Initiating collaboration with one individual at an institution automatically placed me in conflict of interest with *all* grant applications submitted from the entire institution. However, the oversight process also enables the federal government to control the flow of information. In particular, the federal government seeks to avoid all appearances of *promoting a homosexual agenda*.

9. This quote appears in Pattatucci-Aragón A (2001). Contra la Corriente (Against the Current). In: Gartrell NK and Rothblum ED (eds.), *Everyday Mutinies: Funding Lesbian Activism*. Binghamton, NY: Harrington Park Press, p. 8.

10. Heteronormativity describes a binary gender system in which only two sexes are recognized, where sex is equated with gender and gender with a heterosexual orientation.

11. Restrictions on marriage have a long and disgraceful history in the U.S. Those pertaining to opposite-sex couples historically have centered on race and mental status. For approximately the first 100 years of U.S. existence, those of African descent were by law defined as 4/5 of a human being. While the abolition of slavery struck down this legal definition explicitly, it was implicitly applied for nearly another century. Laws prohibiting interracial marriage, more broadly applied to all people of color, existed in

some states well into the 1960s. The most common justification for maintaining these laws was the erroneous belief that interracial marriages would lead to the birth of monster-like children. The subtext of this claim is that people of color are considered less than human. (For more information see Klarman MJ (2004). *From Jim Crow to Civil Rights: The Supreme Court and the Struggle for Racial Equality.* New York: Oxford University Press.) In another example, the eugenics movement in the first half of the 20th century led to the creation of a subhuman class known as the *feebleminded.* The justification for restricting marriage in this case was that if the feebleminded were allowed to reproduce, *the condition* would contaminate the gene pool of the White majority and spread throughout U.S. society like a plague, resulting in decreased intelligence and eventually leading to the collapse of the country. In addition to the absurdity of this claim, a major problem was that definitions of feeblemindedness were subjective. Anyone could be labeled feebleminded and hauled off to an institution. (For further information see: Black E (2003). *War Against the Weak: Eugenics and America's Campaign to Create a Master Race.* New York: Four Walls Eight Windows.) Although laws prohibiting the feebleminded from marrying have been repealed in all states, policies regarding marriage for the mentally retarded and individuals with acute psychiatric illness remain in some states. Again, the assumption behind the prohibition is that of a subhuman status, in this case a presumed incapability to control sexual urges and remain faithful to a spouse, along with the inability to understand and appreciate the sanctity of marriage. Given this history, one could argue that the heteronormative majority considers lesbians and gay men to be 4/5 of a human being or some other fraction thereof and thus feel justified in prohibiting same-sex marriage.

12. For a more thorough conceptualization of this element, see Kelly Coogan's article featured in this volume: Fleshy Specificity: (Re)considering Transsexual Subjects in Lesbian Communities.

13. Rejection of patriarchally defined gender-role standards became codified for lesbians during the gay liberation movement of the late 1960s, when they broke away early in the process and aligned themselves with the women's liberation movement. Consequently, the politics of oppression displaced romance and love as the central framework for lesbian identity. Once prized in lesbian circles, femmes were vilified (at least publicly) as reifying the patriarchal oppression that *real lesbians* were fighting to end (see Nestle J (1992). *The Persistent Desire: A Femme-Butch Reader.* San Francisco, CA: Alyson Publications). Given this backdrop, the love/hate relationship that lesbian communities have had with femmes can be best understood by the fact that for many lesbians their entry into lesbian communities involved a process of self-discovery and empowerment through the systematic unlearning or outright rejection of patriarchal definitions of femininity and expectations for acceptable feminine behavior. I submit that this path to lesbianism is so engrained in modern lesbian politics that it would be difficult, if not inconceivable, for many lesbians to accept that a femme woman could arrive at her lesbian identity through a different route, a route that involves *embracing* the same patriarchal standards for feminine appearance and behavior that others rejected to establish their lesbian identity. Because the embracing route is considered a less noble pathway to lesbianism than rejection, femmes occupy a minority status within lesbian communities. Femme women tend to be seen as less genuine lesbians and are viewed with constant suspicion (e.g., that they only are interested in sexual play and might run off with a man at any moment). However, if we conceptualize patriarchally defined gender-role standards as a *uniform*–a set of symbolic signifiers that denote submission and sexual availability to men that women are required to

wear–then which lesbian, the one who rejects the uniform or the one who embraces it, is engaged in the most radical act? I propose that wearing the uniform but not for its intended purpose is a greater act of in-your-face radical subversion than rejecting the uniform entirely. The love/hate relationship with femmes can be further understood using the antiquated adage *butch in the streets, femme in the sheets* as a metaphor. Femmes may be disparaged in public spheres, where lesbian values take on their greatest importance, but they still are desired privately. A recent poll on the Website for the television series *The L Word* confirms this. Producers of the show asked fans to indicate what they would like to see more of for the second season of the show. Among the various categories, *more gorgeous femmes* was the overwhelming favorite. This is a case where internal desire, expressed within the context of an anonymous poll, trumped external group-driven politically correct behavior. As Judith Butler articulates, patriarchally defined gender role standards congeal on the surface to generate the appearance of an internal core identity (see Butler J (1990). *Gender Trouble: Feminism and the Subversion of Identity*. New York: Routledge, p 136). Although lesbian communities act as if the process is (or should be) different for lesbians, in reality it is not. Gender role standards in lesbian communities also congeal on the surface to create the appearance of an internal core identity. These standards are normalized and are just as zealously policed as those observed in mainstream heterosexual settings. In this respect, *homonormativity*, like its sister *heteronormativity*, is a value-laden façade projecting an ideal–how things should be or how we wish them to be.

14. Although same-sex marriage is legal in Massachusetts, I submit that the general public tends to view same-sex unions as a lower form of marriage or as entirely invalid. This is evidenced by fervent attempts in Massachusetts to amend the state constitution to define marriage as a union between one woman and one man and equally ardent attempts to amend the U.S. Constitution to define marriage once and for all. Civil unions and domestic partner benefits by definition are considered a *lower species* of marriage by the heteronormative majority. Thus, the heteronormative standard of marriage remains firmly in place as the desired form or the ideal, despite the fact that some marriage or marriage-like rights have been won for same-sex couples. However, most important for this discussion is the way in which homonormativity acts to fragment lesbian communities into hierarchies of worthiness. Lesbians that are able to come closest to mimicking *heteronormative* standards (we're just like you except for one thing) are deemed the most worthy of receiving rights. Those at the bottom of the hierarchy (intersex, trans, intersectional, and queer individuals) are seen as an impediment to this elite class of lesbians obtaining their rights. This is evidenced by the unfortunate, but all too common, tendency among homonormative lesbians to blame the victim when trans persons suffer violent attacks or are murdered. On the surface, these lesbians seem to be asserting that if the trans person were a little less trans, a little less flaming and a little more closeted, they would not have been victimized. However, imbedded in the subtext is a frustration over the necessity for damage control with the heteronormative majority (e.g., Don't freak out. Remember, we're like you; not like them).

15. For an introduction to Social Dominance Theory, see Sidanius J and Pratto F (2001). *Social Dominance: An Intergroup Theory of Social Hierarchy and Oppression*. Cambridge, UK: Cambridge University Press.

16. I am indebted to Kelly Coogan at Rutgers University for this quote and for helping me to flesh out concepts in this introduction.

Fleshy Specificity:
(Re)considering Transsexual Subjects
in Lesbian Communities

Kelly Coogan

SUMMARY. The transsexual subject position has recently been subsumed under the labeling "transgender." This article considers the implications of this new labeling for transsexual subjects and their capacity to be incorporated into, or at least configured as affiliated with, lesbian communities and their popularization of transgender terminology. By tracing the complicated academic and political etymology of "transgender," this article shows how the term unfairly erases the lived experi-

Kelly Coogan recently graduated from Duke University with a degree in Women's Studies and a Certificate in Sexuality Studies. As an undergraduate, she completed an honors thesis that pushed poststructuralist feminist theory into a more sustained and legitimizing conversation with biology. Her thesis research led to one of her current intellectual interests in transsexuality and its ambivalent relationship to feminist theory. At present, Kelly is a graduate student in the Department of Women's and Gender Studies at Rutgers University. Her dissertation will focus on authoring different genealogies of Women's Studies, as an institutional field formation and as a (post)identitarian and interdisciplinary knowledge project, in the U.S. and transnationally.

Address correspondence to: Kelly Coogan, Women's and Gender Studies, Rutgers University, 162 Ryders Lane, Second Floor, New Brunswick, NJ 08901 (E-mail: kellcoog@rci.rutgers.edu).

[Haworth co-indexing entry note]: "Fleshy Specificity: (Re)considering Transsexual Subjects in Lesbian Communities." Coogan, Kelly. Co-published simultaneously in *Journal of Lesbian Studies* (Harrington Park Press, an imprint of The Haworth Press, Inc.) Vol. 10, No. 1/2, 2006, pp. 17-41; and: *Challenging Lesbian Norms: Intersex, Transgender, Intersectional, and Queer Perspectives* (ed: Angela Pattatucci Aragón) Harrington Park Press, an imprint of The Haworth Press, Inc., 2006, pp. 17-41. Single or multiple copies of this article are available for a fee from The Haworth Document Delivery Service [1-800-HAWORTH. 9:00 a.m. - 5:00 p.m. (EST). E-mail address: docdelivery@haworthpress.com].

Available online at http://www.haworthpress.com/web/JLS
doi:10.1300/J155v10n01_02

ences of transsexual subjects by ignoring their specificities in the flesh. The article concludes by offering alternative disciplinary matrices and conceptual schemas through which to better imagine transsexual subject formation and transsexual subjectivity. *[Article copies available for a fee from The Haworth Document Delivery Service: 1-800-HAWORTH. E-mail address: <docdelivery@haworthpress.com> Website: <http://www.HaworthPress.com> © 2006 by The Haworth Press, Inc. All rights reserved.]*

KEYWORDS. Bodily materiality and biology, Brandon Teena, Butch/FTM Border Wars, gender transitioning, feminist theory, queer theory, linguistic performativity, theatrical performativity, sexed ontology, sex transitioning, transgender and feminism, transsexualism and lesbian theory, transsexualism and feminism, transvestitism and feminist theory

Flesh so named makes difference.

–Jacquelyn N. Zita (1998, p 106)

In 1998, *GLQ* published a special *Transgender Issue* where Judith Halberstam and C. Jacob Hale first discuss the trope of the *Butch/FTM Border Wars* in a collaboratively written introduction to their essays in the issue.[1] Citing notions of space, geography, territory, and territoriality as just as central to the politics of gender and sexuality as notions of time and temporality, Halberstam and Hale introduce the trope of the border war in order to capture the intense, yet sometimes subtly articulated, political battles waged by differing marginal subjects for visibility and inclusion within, as well as affiliation with, lesbian communities. Specifically, the terminology *Butch/FTM Border Wars* denotes disputes between butches and female-to-male transsexuals (FTMs) concerning what bodies and identifications are appropriately "butch" or "transsexual." Butches are usually, but certainly not always, more visibly queer in relationship to their masculinity than FTMs. Outside of lesbian communities the general public oftentimes reads the queerness of butches *in an instant*. The visual markings of gender transitivity on their bodies reveal that they inhabit a space in between male and female, or what Halberstam and Hale term the *border-zone*. On the other hand, within lesbian communities butches frequently fit group contours more easily than do transsexuals due to their anatomically sexed female bodies.

My definition of lesbian communities in this essay is somewhat loose and most likely controversial. In my mind, lesbian communities include all those queer bodies which rearticulate heteronormative prescriptions of sex, gender, and sexuality in enormously complex ways but are anatomically female. This is not to say that hierarchies within these communities amongst differing female bodies (some that are considered *more* lesbian than others or consider themselves to be the *best* kind of lesbian) do not exist because they surely do. Rather, I hope to hone in on the fact that many lesbian communities continue to be haunted by a very hard-to-shake *a priori* grounding–that of the anatomical facticity of the female body–despite many nuanced attempts by contemporary feminist, lesbian, and queer theorists to challenge the givenness of female anatomy as apt grounds for lesbian politics and scholarship. The anatomical female body *always* and *already* manages to sneak back in as grounds–meaning that it serves as the political grounding point of lesbian coalition amongst the increased emergence of disparate bodies desiring to be called *lesbian*–even if the female body's grounding as such is radically repudiated on several different post-identity political and theoretical fronts. Nowadays lesbian group contours are increasingly democratic, sometimes even deceptively so. They are loose enough to include a myriad of queer bodily variations, just as long as those bodies are judged anatomically female at the end of the day.

This critique has been rehashed so frequently that it is almost dismissed just as soon as it is staged. But I stage this critique through a different lens, a lens that might shift its terms productively: that of the transsexual subject, *his* or *her* subject formation and *his* or *her* subjectivity. Many lesbian communities criticize and deny transsexuals inclusion, and sometimes even affiliation, due to their ostensible entrance into (hetero)normativity and their capability to *pass*[2] in mainstream culture. FTMs are ousted by such communities due to their supposed arrival into a stable male-sexed ontology via hormone therapy and/or sex-change surgery.[3] Doing away with their anatomically female bodies makes the entrance for FTMs into lesbian communities extremely difficult if not entirely impossible. In fact, one might argue that they habitually serve as the constitutive outside or as the abject of lesbian communities. In the scenario of the butch/FTM border wars, FTMs are in the offensive position, most usually outside lesbian borders, and in some cases fighting desperately for inclusion or at least affiliation.

In its consideration of transsexual subjects, this essay hopes to circumvent rehashing the argument that lesbian communities should try, once again, to do away with the anatomically female body as its ground, a repudiation that supposedly would allow room for FTMs, inter-sexuals, and politically sympathetic men. When lesbian identity politics, and even, as I am arguing, some versions of post-identity politics, rely on the anatomically female body for their political grounding, they are usually cast as being essentialist. Yet, even though essentialism, especially biological versions of it, can be highly dangerous for several different theoretical and political reasons, knee-jerk reactions against essentialism make me suspicious especially when they call for doing away with biological discussions of bodies altogether. Such reactions against essentialism are hardly theoretical and political panaceas. The biological organism of the body somehow always manages to return. It cannot simply be ignored and nor can biological sex, even if sex is no longer seen as mapping out the truth of our bodily destinies.[4] If the case of transsexuality reveals anything at all, it shows us that biological sex is just as *transitory* and *un*natural as gender, but certainly no less relevant in defining our ontology. To ignore sex's role in constituting our sense of who we are is, in many ways, to *re*naturalize and *re*essentialize it. By ignoring sex altogether, in other words, the transitivity and fluidity of biological sex (even if its versions of transitivity and fluidity are different from those of gender) is left thoroughly under-theorized and under-explored by those theorists needing to study it most.

This essay poses the following questions. How would the way lesbians imagine the borders of their communities have to change in order to include transsexuals with more regularity? How could feminist and queer theories quicken this refashioning of boundaries by shifting their own relationship to biological determinism, essentialism, and the elasticity of the body's fleshy contours? What theoretical and methodological moves will feminist scholars need to make such that transsexual subjectivity may be theorized in ways that do not foreclose its conditions of existence? Perhaps a reconsideration of the place of transsexuals in lesbian communities demonstrates that these communities still succumb to the pressures of a biologically female-sexed ontology in defining who counts as lesbian and who does not. . . . Perhaps the specter of essentialism haunts these communities in a profoundly anti-essentialist guise. . . . As Henry S. Rubin so eloquently suggests,

> Nontranssexuals assume the coherent legibility of their gendered embodiment or their identities and are not expected to carry a share of the revolutionary burden of overthrowing gender or imagining what to replace it with. They do not walk around, as they seem to be asking us to do, without gender identities or legible

bodies. No matter how much gender play they engage in, few of them will say that they are not women or men after the clothes come off. They are not called upon to account for the fact that their gender is something they achieved. (1998, p. 273)

Several contemporary feminist theorists deeply fear the idea of sexed ontology, where biological sex is posited as expressing a truth of one's being. According to critics of sexed ontology, to argue that biological sex is ontological is to be a biological essentialist. In other words, when one argues that sex is ontological, one is seen as giving into the facticity of biological sex, or the givenness of the idea that there are two, *and only two*, versions of biological sex. In this scenario where sexed ontology is equated with sexed facticity, one's way of being in the world is determined by the *fact* that one is either biologically male or biologically female. When sexed ontology and sexed facticity are equated, all gendered behaviors supposedly stem from this male/female sexed dichotomy. But is there a way to disentangle this equation of sexed ontology with sexed facticity? What if biological sex is just as complex and multifarious as gender is said to be?[5] If sexed ontology develops from an understanding of sex that is just as fluid as gender, what happens to the givenness of binary sex? Sexed ontology is not *necessarily* the same thing as sexed facticity. Sexed facticity in biology, or the seemingly natural givenness of two biological sexes, is constitutive of *a single* sexed ontology but is indeed only one of ostensibly several.

No doubt transsexuals epitomize an investiture in sexed ontology through their sex transitioning. They wholeheartedly believe that their entrance into a new biological sex will be constitutive of a new way of being in the world and relating to reality. But they hardly epitomize sex's facticity in biology. If anything, their very existence–along with the existence of intersexuals and hermaphrodites–does away with the idea that two *and only two* biological sexes are naturally given. The subtleties, nuances, and radical differences of the distinction between sexed ontology and sexed facticity are decidedly under-articulated and must be further explored.

I propose that lesbian communities could include transsexual subjects in a way that is sensitive to their everyday lived experiences by doing two things: (1) admitting to the transitivity of biological sex and (2) articulating how the transitory character of sex is both similar to and different from the transitory character of gender. It is the task of specifying versions of the transitory character of sex–of specifying the permeability of transitioning that occurs within the register of the flesh in transsexual subject formation *and* in the formation of transsexual subjectivity–that lesbian, feminist, and queer theorists[6] must engage in their scholarly practices and lesbian communities must engage in their political practices. In my mind, these constituencies–academic lesbians, queers, and feminists as well as lesbian communities–participate in the production of imaginative practices of inclusion, or *imagined communities* as Madelyn Detloff talks about,[7] which either work to include or exclude (sometimes inadvertently) transsexual subjects. This essay is concerned with identifying some of these imaginative practices and more critically discerning how they work and their unintended material consequences or affects for transsexual subjects. The essay started with the trope of the border war because it very nicely exposes how transsexuals and their subjectivities might easily be overlooked by the aforementioned constituencies. As Halberstam has noted, the trope falsely situates butches and FTMs along a masculine continuum, where butches are on the far left, as least masculine, and transsexuals are on the far right, as literalizing masculinity. Although the continuum purportedly represents the multifarious ways specific bodies embody masculinity, it fails oftentimes to properly denote the *highly specific* ways that some subjects feel disembodied by their masculinity and then, in turn, work towards masculine *re*embodiment.[8] I argue that this representational failure is especially true for FTMs.

Despite their polar placement along this continuum, *butches* and *FTMs* are now seen as justly falling under a new popular label, *transgender*, a labeling that fails to fully capture the specific bodily (re)mappings each undergoes. Transgender currently acts as an umbrella term for a campy assemblage of bodies and identities, like those posited by Susan Stryker such as "FTM, MTF, eonist, invert, androgyne, butch, femme, nellie, queen, third sex, hermaphrodite, tomboy, sissy, drag king, female impersonator, she-male, he-she, boy-dyke, girl-fag, transsexual, transvestite, transgender, cross-dresser" (1998, p. 148).[9] Since the term has boundless elasticity, several lesbian subjects undoubtedly use it as a self-descriptor. Identifying as transgender, or at least as affiliated with the category of transgender, places these lesbian subjects in coalition with other lesbians who articulate their queerness in profoundly divergent ways. With Stryker's comprehensive list, it is easy to see how the term transgender performs two very important functions for lesbian, queer, and feminist scholars in assisting lesbian communities to better imagine their contours. First, transgender acts as a term which purportedly *represents* countless queer bodies. It just about allows for transsexual inclusion but at the expense of erasing the specificity of their lived experiences of transitioning as I discuss below. This erasure of their specificity says to transsexuals regarding their inclusion in lesbian communities: *Almost but not quite*–you can kind of be included, but within the terms and conditions of our understandings of the transitory character of sex and gender and not yours. You must narrate your sense of being transitional according to our understanding of being transitional. Second, in its sustained focus on the descriptive, hermeneutic, and analytical potentials of *(sex)uality*, queer theory has caused ruptures in the theorization of gender by contemporary feminist theory, a phenomenon I will also elaborate below. Transgender attempts to very strategically *bridge* the gap between feminist and queer theories and thus carries the potential to heal these ruptures. With transgender, feminist theorists who take seriously feminism's engagement with queer theory wipe off their sweaty brows in relief, as the term inserts gender and its transitoriness as absolutely central to the process of queering. Yet, a presupposition of the term transgender is that seemingly any *body*, irrespective of its anatomically assigned sex at *birth*, can circulate *freely* through the *trans*, arriving at some gender identifications and leaving again *momentarily* for others.[10] No differentiations in the transitioning process are inherent in the term. For instance, the transitioning of a cross-dresser is different from that of a transsexual, even if biological sex change, with modern biotechnology, is rendered representationally similar to the changing of one's clothes. Such transitionings might be analogous but they are certainly not the same, yet both a cross-dresser and a transsexual might be seen as justifiably transgender. And both might even want to identify themselves as such.

In an effort to consider the specific lived experiences of transsexuals, I must bring the *trans* celebration of free-floating gender signifiers to a screeching halt. If the theoretical and political hold on transgender is driven by an impulse to include the myriad bodies that wholeheartedly wish to fall under its rubric, will not the term invariably fail to display such endless inclusive capacities in the same way that say *women* fails feminist theory? In the rush towards more inclusive terminology, theorists and activists alike oftentimes forget the arbitrary *beginnings* of their inclusive impulse that presumably stemmed from a desire to represent the specific lived experiences of sexually and gendered marginal subjects with increased accuracy, whether those experiences cohere around pleasure or pain. While umbrella terms such as transgender enable us to imagine politics with augmented inclusion and affiliation, they also risk rendering invisible the precise lived experiences of different bodies *in transition*. For instance, the term transgender risks obscuring the pains and sufferings of subjects for whom sex and gender transitioning is a matter of life and death, like many transsexuals. It also risks eliding the different temporalities of sex and gender transitioning for different subjects. Some versions of transition take minutes and others take years. No version of transition is necessarily more or less real than the other, but all experiences of transitioning deserve to be told, despite their differences.

In no way do I wish to suggest that every subject identifying as transgender unabashedly participates in a gender play that drips with hedonism. Nor do I wish to convey that the ontological condition of transsexuality (not that there is only one) is characterized entirely by long drawn out pain and suffering. Transgenders undeniably experience considerable amounts of pain and suffering at times, and transsexuals undeniably experience considerable degrees of pleasure and satisfaction. The definitional boundaries marking transgender and transsexual distinctness are shifting incessantly; nevertheless, they are still distinct. In an era when gender fluidity is in vogue, it is our responsibility to diligently delineate specificities within and amongst particular gender identity categories. This call for specificity is not to delegitimize the transgender subject position for those who continue to identify as such while receiving hormone therapy or other anatomical surgeries in efforts to feel more sexually embodied, or for subjects who are intersexed or hermaphroditic and have had surgeries imposed on them at birth. The space of sexed and gender in-between-ness can be considerably unbearable for many and deserves the same amount of attention as many transsexuals' desired space of sexed stability does. The transgender experience is not *essentially* pleasurable and the transsexual one *essentially* painful, each are attendant with various pleasures and pains, nor are transsexuals the only subjects seeking anatomical transition. Nonetheless, I am interested in opening up conceptual space for lesbian communities to seriously consider the specific lived experiences of those subjects who transition in the flesh, who shift their bodies' fleshy contours due to perceived sexed disembodiment. These subjects are too often subsumed under transgender or not considered at any length due to their belief in a literalizing fantasy of sex. While many subjects identifying as transgender daily live the pains of their liminality, the *representational matrix* of transgender most often covers over gender's sufferings by conceptually privileging the fluid and reiterative pleasures of gender transitioning, a supposed antiontological space of gender play that, in reality, comes with its own set of ontological commitments around gender's inherently fluid capacities.

The representational matrix of transgender not only effaces the specificities of the various bodies it purports to describe, but it also erases the deeply complicated and contestatory relationship transsexual subjects have had with feminist theory and politics since the late seventies.[11] This cleavage occurs most frequently in those versions of feminism that privilege the subjectivity of lesbian subjects who are anatomically female and find their pleasure, empowerment, and politicization via the sexed ontology of their bodies. If theory and practice are mutually constitutive, feminist theorization of transsexuals as politically adverse to or invisible within the category of lesbian significantly marginalizes and excludes an extremely diverse constituency of sexual minorities. From Janice Raymond's 1979 publication *The Transsexual Empire: The Making of the She-Male*, to Judith Halberstam's infamous injunction that "There are no transsexuals. We are all transsexuals," the transsexual subject is either radically assaulted by lesbian theorists or swallowed up in the complicated gender and sexual play of postmodernity and postmodern predications of lesbian identity formation.[12] Akin to Halberstam's initial formulation, some postmodern theorists have figured transsexuality as *the* paradigmatic bodily aesthetic of postmodernism, whereby everyone is figured to slide into and out of certain sex and gender aesthetics and practices through time and in different spaces.[13] In either account–the lesbian feminist or postmodern–the transsexual desire for sexed transitioning in the *flesh* is lost without a trace.

A PROFOUND SHIFT IN IDENTITARIAN KNOWLEDGE FORMATIONS WITHIN THE ACADEMY

Why is the border war a descriptively and analytically useful trope for feminist and queer theorists to study the battles over identity and identification within lesbian communities? Perhaps because it aptly encapsulates the hard-fought struggles for visibility and comfortability, laced with intense pain, which mark the perceived need of certain queer constituencies who feel marginalized within their own communities.[14] It provides representation, albeit only ever partial, to the lived pain of being queer and existing in queer communities amongst other queers, fighting for scarce resources and limited recognition. It identifies the reality that to be *gender-transit* can be either a privilege or an imposition. Gender transitioning can mark sexed and gendered play, and it can also denote the stark reality that an FTM cannot afford sex-change surgery and must turn tricks in order to save up enough money to one day transition. The specificities of the transitional character of gender have only recently begun to be drawn out thanks to Halberstam's and Hale's groundbreaking formulation of the border war trope. But how did we arrive at this lack of specificity in the first place?

If I were performing an archaeological tracking of how identitarian knowledge formations profoundly shifted in the academy, I would select U.S. feminist and gay and lesbian (not yet queer) theories in 1990 as one point of considerable archaeological shift and, thereby, an apt point of departure. Two major texts were published that year: Judith Butler's *Gender Trouble* and Eve Kosofsky Sedgwick's *Epistemology of the Closet*. Tirelessly poked, prodded, and hailed, these texts no doubt significantly shifted the terrain of theorizing gender and (sex)uality in the academy. Both theorists wrote at a time when their work was direly needed. Being submerged in poststructuralist and postmodernist analysis, feminist scholars desperately desired a non-volitional account of gender identity formation that Butler provided. Likewise, gay and lesbian theorists were frantically pining for a rigorous theorization of sexuality distinct from gender, an analysis Sedgwick devotedly worked towards.[15]

To simplify for brevity's sake, Butler's theory of gender performativity warrants that psychoanalysis, in its attempt to articulate the organizing principles of identity, augments the complicated social, cultural, discursive, and political origins of gender identity onto a psychological core. She argues that this displacement comprehends "words, acts, gestures and desire" to "produce the effect of an internal core or substance" on the surface of the body (1990, p. 136). The surface of the body is the site for a *literalizing* fantasy of stable sex, as the organizing principles of identity congeal to reveal the form of a stable sexed ontology on the body's surface. Working against the fantasy that sex expresses the truth of one's being, the theory of gender performativity undoes any psychoanalytic notion that the body congeals into a finalized form once identity is considered fully organized, and it articulates the congealment of the body's surface to the extent that it comes to signify the facticity of sex.[16] Butler reworks female masquerade, the gender ontology of Lacan, and turns Freud's melancholia into a gender ontology so as to conceive of the body's sexing as dependent upon the elasticity of the body's surface, which is rearticulatable incessantly through an individual's acts of loving, desiring, losing, and incorporating those lost objects of love and desire *ad infinitum*.[17] Gender and sexuality are inextricably intertwined in her account.

Sedgwick, on the other hand, works hard to disarticulate sexuality from gender by reading the Western European English canon for its perversities and self-differences. Perhaps the most read introduction in the history of Gay and Lesbian and Sexuality Studies, "Axiomatic" lays out seven central axioms for the study of sexuality. Number two is most relevant here: *"The study of sexuality is not coextensive with the study of gender; correspondingly, antihomophobic inquiry is not coextensive with feminist inquiry. But we can't know in advance how they will be different"* (Sedgwick, 1990, p. 27). For Sedgwick, when talking about sexuality, an analytic of gender difference only gets at the gender of one's sexual object choice. It fails to get at "the array of acts, expectations, narratives, pleasures, identity-formations, and knowledges, in both women and men, that tends to cluster most densely around certain genital sensations but is not adequately defined by them" (Sedgwick, 1990, p. 29). While Sedgwick set Gay and Lesbian and Sexuality Studies on intellectual fire, in feminist circles not everyone was overjoyed by the proliferating play of sexual signifiers and practices.

The queer break from feminist inquiry figured academic feminism as anachronistic, as the *has-been* of the intellectual avant-garde.[18] Playing catch-up, some factions of feminist scholarship began increasingly to interact with poststructuralism and its repudiation of referential theories of meaning, forming a body of scholarship now known as poststructuralist feminism, which dispelled any notion that sex has a lasting, factual hold on gender. The marriage of certain gender referents to particular bodies was fast ending in divorce, creating conceptual space for seemingly free-floating signifiers to arbitrarily attach themselves to some bodies and disconnect themselves from others. Supporting sexed ontology was way out of style. Buying into gender ontology was for those who did not have the apt cultural capital for gender ambiguity. Antiontology was now in vogue and so were antiontological epistemologies, hence the overemphasis on epistemology in poststructuralist feminist scholarship and queer theory. The free-floating sex and gender referents touted by poststructuralist feminism began to bode well with the proliferation of pleasurable behaviors and desires that queer theory holds in such high esteem. *Women* no longer fit the bill for the apt subject position of feminism in its convergence with queer theory. A new and more fitting subject position was burgeoning.

Three years later, that subject position–transgender–appeared first for lesbian communities in order to capture a quite different experience than the gender-transit one touted by feminist and queer theorists in the academy. It was a position characterized by the profound pain and loss in gender identification rather than its potential for proliferating pleasures and desires. On Christmas day 1993 a twenty-one-year-old anatomically sexed female who dressed and frequently identified as a male was kidnapped, made to expose his vagina, and forcibly raped by John Lotter and Marvin Thomas Nissen. A week later on December 31, 1993, he was murdered by these men in a farmhouse in Humboldt, Nebraska. When Lotter and Nissen were finally convicted in February 1996, *Saturday Night Live*'s Norm MacDonald made the following horrid comment during his *Weekend Update* segment, "In Nebraska, a man was sentenced for killing a female cross-dresser, who had accused him of rape, and two of her friends. Excuse me if this sounds harsh but, in my mind, they all deserved to die."[19]

His legal name was Teena Renae Brandon (the name that appeared on his birth certificate as well as his gravestone), but, as C. Jacob Hale documents, he went by a whole assemblage of masculine and gender neutral names throughout his short life such as "'Charles Brandon,' 'Brandon Brinson,' 'Ten-a Brandon,' 'Billy Brandon,' 'Brandon Brayman,' 'Tenor Ray Brandon,' and 'Charles Brayman'" (1998, p. 312). Hale shows how *Brandon Teena* is a false stabilization of the victim's name utilized by various lesbian and transsexual constituencies to demonstrate the very real violence enacted on their identities. According to him, the name Brandon Teena developed out of a border war between lesbian journalist Donna Minkowitz who wrote an article in the *Village Voice* entitled, "Love Hurts. Brandon Teena Was a Woman Who Lived and Loved as a Man," claiming that s/he was a lesbian, and the famous transsexual activist group Transsexual Menace who vociferously picketed such claiming, as they thought he was justifiably part of their community. The conditions of Brandon Teena's possibility as a political subject post-death (if there is such a thing) grew out of his naming by queer communities who attempted to secure his traumatic existence and extremely violent death as their own conceptual device to represent the violence in queerness, a stabilization that in the end is unfair to the marginal subject who died that day. A term, a subject position, was needed to describe experiences of subjects like Brandon Teena. Transgender seemed apt to perform this descriptive work as the *trans* promised to capture the pain and violence in sexual and gender ambiguity. In the academy around the same time, though, the term was being appropriated for much different ends not at all connected to the initial political impulse of lesbian and transsexual communities to describe the violences of gender marginality.

When the hype around Brandon Teena's atrocious murder was at its peak, in the mid-nineties, scholars were quickly finding ways to regender queer and rearticulate gender and sexuality. Books describing the transitional character of gender through the terminology of transgender emerged to denote the multifarious ways individuals could *wear* their gender differently–*queer* their gender–from one day to the next.[20] Kate Bornstein's *Gender Outlaw* and Marjorie Garber's *Vested Interests* are two such books that situate bodies as cultural texts which are rendered readable as queer through their regendering or their *trans*gendering. By asking her readers questions about their gender identity in a manner ringing of postmodern pastiche, Bornstein situates gender along a volitional performative schema.

- What's your gender?
- When did you decide that?
- How much say do you have in your gender?
- Is there anything about your gender or gender role that you don't like, or that gets in your way?
- Are there one or two qualities about gender that are appealing to you, enough so that you'd like to incorporate those qualities in your daily life (1994, p. 96)?

In a similar vein, Garber reads cross-dressing, or transvestitism, as a cultural practice of regendering that disrupts Western binary thinking by introducing a third position into the male/female dyad. Cross-dressing, in upsetting the normative-sexed readability of bodies, destabilizes the binary-ness of categories that are *always already* sexed. The liminality of cross-dressing produces a modality of articulation outside of male or female, offsetting the binarism's cultural, social, and historical power.

> The "third" is that which questions binary thinking and introduces crisis–a crisis which is symptomatized by *both* the overestimation *and* the underestimation of cross-dressing. But what is crucial here–and I can hardly underscore this strongly enough–is that the "third term" is *not* a term. Much less is it a *sex*, certainly not an instantiated "blurred" sex as signified by a term like "androgyne" or "hermaphrodite," although these words have a culturally specific significance at certain historical moments. The "third" is a mode of articulation, a way of describing a space of possibility.

Three puts in question the idea of one: of identity, self-sufficiency, self-knowledge. (1992, p. 11)

According to Garber's reading of transvestitism, queer expressions of one's gender are as simple as changing one's dress, putting on and taking off certain clothes, as she disarticulates gender from anatomy so that sex purportedly becomes readable through dress, enabling queer readings of the body's *clothed* surface. For Garber, gender's intelligibility is performative in the *theatrical* sense, not in the *linguistic* sense, a distinction that too often gets quickly glossed over. Whereas Garber sees the transvestite as putting on and taking off clothes with intention, in Butler's theory of gender performativity, by contrast, gender appears on the surface of the body through unconscious, nonvolitional loss. Both Bornstein's and Garber's understandings of gender's versatility, readability, and wearability provide context for statements like those made by a young Judith Halberstam that "There are no transsexuals. We are all transsexuals" (1994, p. 212) as if the body's sexing is akin to putting on and taking off clothes at will. After all, if anatomical sex no longer has any realness, why wouldn't we all be transsexuals? If skin can just as easily be taken off and put back on as changing our clothing (and it of course cannot) perhaps the transsexual is no longer a needed possibility? Sex and gender are embodied along profoundly different temporal registers when one cross-dresses than when one receives hormone therapy or sex change surgery. In the latter case, a transsexual must *live* each day in *his* or *her* newly sexed body. A cross-dresser may feign maleness for an evening or even several evenings in a row and still not access, or feel the need to access, the same degree of sexed ontology a transsexual often strives for. Uncritically aligning transsexual lived experiences with those of cross-dressers or other gender-queer bodies not shifting their bodies' biological contours unfairly erases fleshy specificity.

For Bornstein and Garber, regendering queer is a volitional practice of reading bodies in certain non-normative ways. Their gender politics are processes of *(re)aestheticizing bodies*, resignifying sexual anatomy with dress and other bodily surface manipulations. Dress becomes one's sexual skin, and transsexuals, through sex-change surgery, are seen as merely refiguring one version of their body's dress code, that of the skin, amongst many. Regendering queer denotes the versatile means through which the body's surface can endlessly resignify sex in ways arbitrarily related to one's sexual object choice. By disconnecting sex from anatomy, Garber and Bornstein (and even Butler at times) imagine they are antiontologizing sex entirely. It is within this academic context that the *trans* prefix garnered a pleasurable, endlessly inclusive valence, straying far afield from its painful birthing via Brandon Teena's death. *Trans* currently acts as a prefix *reborn* to capture the pleasures, desires, and occasional pains (most usually psychical) that subjects ostensibly experience now that sex (the signified) and gender (the signifier) are subsumed into one another, or more specifically when sex is cast as the discursive affect of gender. *Trans*, in other words, attempts to clean up the referential mess resulting from antiontologizing sexed bodies. An entire set of bodies and identities easily fit under transgender, the new women of poststructuralist feminism.[21]

The political project of antiontologizing sex *is* a gendered (re)aestheticization of bodies and *is* a politics working primarily within the register of bodily surfaces. *The antiontologizing gesture risks rendering bodily surfaces–their reaestheticization, resignification, remapping, etc.–the entire terrain of the political, i.e., everything, in poststructuralist feminist and queer politics.* As I see it, there are two ways to redress this overemphasis on surfaces. The first would be to more finely tune and intricately layer the antiontologizing gesture in its concerns and effects, although it is hard to say in advance exactly what such fine tuning and intricate layering would look like or entail. The antiontologizing gesture currently runs the risk of too facilely producing volitional transgender subjects. In other words, with intention, transgenders seemingly pick up, take on, and put down different gender identities at the drop of a hat. As Jacquelyn N. Zita notes, "Postmodernism supplies a set of ontological commitments needed for a world in which the body appears to be malleable, protean, and constructed through and within discourse" (1998, p. 105). These ontological commitments *do not* assume a centered subject, rife with intention in her gender and sexual acts. The protean self that Zita speaks of is not the conscious agent, the cogito of the Enlightenment, whose gendering precedes its predication by and within discourse. Yet theorists like Garber and Bornstein place their *trans*gender subject in the most beneficial and agentic position possible: within this Enlightenment tradition while simultaneously reaping the fluid, transitory, and pastiche-like benefits of its intellectual offspring, postmodernism. No doubt any transgender subject emerging from this context would have a pleasurable existence. The second way this overemphasis on surfaces can be redressed is through *re*ontologizing[22] sex *differentially*–that is, asking what it means politically and theoretically to subscribe to a sexed ontology that is not mired in a dualistic account of biological sex but is rather in tune with sex's inherent biological complexities.

MORE THAN WHAT APPEARS ON THE SURFACE

In the past decade and a half, feminist scholars have overtly problematized volitional accounts of gender acquisition. The transgender subject implicitly posited by Bornstein and Garber fetishizes the manifest *consciousness* of gender's performativity, a consciousness no longer seen to exist outside of or in excess of its emergence in discourse. This notion of a transgender subject who fashions *his* or *her* gender at will for a finite period of time fits better within a theatrical theory of performativity, where performance is consciously taken up, not a linguistic one. It is important to note, though, that neither understanding of performance–theatrical or linguistic–aptly captures the experiences of transsexual subjects, their subjectivity or gender acquisition. For instance, in terms of theatrical performativity, the transsexual is hardly an overtly conscious agent who manipulates *his* or *her* fleshy contours to apperceive gender's failure to fully congeal. Transsexuality is a diagnostic category and subjects desiring sex-change surgery must go through myriad medico-juridical institutions prior to receiving permission for change. Transitioning oftentimes takes years of treatment and recovery, sometimes inadvertently resulting in death. There is a high price to pay, monetary and otherwise, for transitioning in the flesh, and it is certainly not a fee overly characterized by gratification and delight.

Yet, despite their inability to fit within volitional accounts of gender acquisition, transsexuals do not conform to Butler's theory of gender performativity either, due to their supposed belief in sex's literalness. If, as Butler suggests, sex was *always already* gender anyway, transsexuals are merely sold short by the medical establishment's emphasis on the truth of sex and really are the cultural dupes many accuse them of being.[23] But Butler shows that sex is *always already* gender via a highly complicated, and sometimes problematic, set of moves that link the formation of the body's materiality to the psychical formation of subjectivity. This linkage proposes a notion of the body's materiality that connects the body to the psyche in such a way that inquiries into the self's psychic structuration risk being seen as addressing the construction of bodily materiality entirely. For Butler, bodily surface signifies the complicated psychic processes whereby psychoanalytic conflicts that occur within the body's interior congeal to produce the body's exterior image. The production of the body's surface appearance is located in these infinite psychic mimetic dramas, and the means through which a subject comes to know her bodily materiality at all is through this phantasmic projection of the body. A subject's bodily matter does not exist prior to its surface projection. This is precisely the point in Butler's theory where transsexuals become misfits, as they are invested in a notion of bodily matter prior to its surface projection. Such a notion explains their feelings of sexed disembodiment and contains the hope that if they do change sex, they will feel more sexually embodied.

Part of this reluctance to conceptualize bodily materiality separate from or prior to one's bodily surface projection is due to poststructuralist feminism's ambivalent relationship with the biological sciences, an ambivalence that has significant historical weight. Poststructuralist feminists understand the biological sciences to be an intellectual project continually reproducing truth claims regarding biological sex, sexuality, gender, and the body. They see the following equation as an inevitable epistemological implication of biological sciences' devotion to studying biological sex: bodies = sexed ontology = sex's binary facticity. But if anything, transsexuals demonstrate that despite their desire for sexed ontology, with advances made in the biological sciences, particular sexed ontologies are not inescapably connected to specific bodies. No longer is any body *forever* wedded to a single sexed ontology, and biological sex does not have the over-determined factual hold on bodies that it once did. Stripped of its sexually factual baggage, bodily reality in the biological sense should emerge as a legitimate site of inquiry for feminism. When it does, feminists will be able to theorize transsexuals and their specificity with increased rigor and sensitivity.

To the extent that poststructuralist feminists consistently relegate claims regarding bodily biological materiality to the realm of essentialism, they are implicitly rendering the body as a biological organism static and are, in effect, re-essentializing it.[24] They are relying on an understanding of the biological sciences that conceives of biology's relationship to bodies as one which is inherently deterministic. As we begin the twenty-first century, the biological sciences conceive of bodies as anything but static. New technologies that rework them emerge nearly every day. Much poststructuralist feminist theory relies on a perception of early to mid-twentieth century biology, which saw bodies as more static and impermeable. No longer a discipline of mere static empirical observation of bodies, the biological sciences of the twenty-first century is a discipline which focuses on bodily reconstruction and rebuilding. At this historical moment, poststructuralist feminist theory should reassess its contestatory relationship to the biological sciences, not to uncritically valorize them, but rather to engage their reconstructive possibilities and the theoretical openings they create.[25]

Transsexual photographer Loren Cameron recently published a book of transsexual photographs entitled *Body Alchemy: Transsexual Portraits*. Using riveting personal narratives and photographs of himself as well as other FTM transsexuals from differing class, racial, and ethnic backgrounds, Cameron takes his reader through the non-linear and never-ending process that characterizes transsexual identity formation. Intertwining narratives and pictures of self-loathing, self-doubt, and self-actualization that occur at different moments pre- and post-surgery, and focusing on the physical bodily changes that facilitate and accompany these shifts in emotional states, Cameron gets at a level of transsexual agency not yet fully articulated: the level of transition that occurs in the flesh. In an emotionally charged self-portrait entitled "Testosterone," Cameron, wearing an extremely gruff face, violently smashes a glass bottle against a chain fence. The text alongside the frustrated portrait reads:

> I inject myself with a dose of testosterone every two weeks, the standard maintenance schedule for men like me. Between injections, the oil-based drug absorbs slowly through the muscle tissue. I admit I've become very attached to taking the hormone, which is responsible for all my physical masculine attributes, like my facial and body hair and muscle development. I've noticed that it affects my sex drive and emotional state too. During the peak part of my cycle, I turn into a randy, greasy kind of guy who is more than a little irritable. (Cameron, 1996, p. 20)

His clever title, *Body Alchemy*, marks an obvious attempt to name the level of transsexual agency that occurs within the biological structures of the body, as alchemy refers to the magical transformation of something common into something precious. As I see it, lesbians have little reason, other than an aversion to biology, to place transsexuals on the fringes of their communities or exclude them entirely, as even an aversion to biology is, to a large extent, unfounded if transsexuals are absolved on the basis of their transition into anatomical manhood. Such exclusion rests on the sexed realness of female anatomy, an implicitly biological assumption in itself. Lesbians should view transsexuals as politically aligned with them in their desires to destabilize patriarchal notions of sex's dualistic facticity and garner some semblance of comfort in their own sexed ontologies. Sociocultural, political, historical, discursive, phenomenological, and especially biological inquiries into transsexual fleshy specificity within lesbian communities mark feminist and queer theorists' next scholarly venues from which to know the *edginess* of transsexuals and their subjectivity.

AUTHOR NOTE

I am enormously grateful to Mary Hawkesworth for reading several earlier versions of this essay and providing me with invaluable commentary and suggestions. Louisa Schein, Elizabeth Grosz, and Angela Pattatucci Aragón, as well as other contributors to this volume also read drafts and imparted very helpful advice, the traces of which are no doubt found within my revisions.

NOTES

1. Halberstam, Judith and C. Jacob Hale. (1998). "Butch/FTM Border Wars: A Note on Collaboration." *GLQ. The Transgender Issue*. 4(2). Durham: Duke University Press, pp. 287–309.

2. Much interesting commentary has been produced regarding "passing." Biddy Martin, in "Sexualities Without Genders and Other Queer Utopias," argues that femme lesbians pass in the same way that many transsexuals do. She states, presumably of her own identification, "I prefer Lisa Duggan's lament, her wish that she were a lesbian drag queen, so that her femmeness might be representable as something other than passing, by way of some sort of routing through a gay male form and back again, a routing that makes the crossings in all forms of identification more evident" (1994, p. 112). Jacquelyn N. Zita criticizes the notion of passing altogether in "Male Lesbians and the Postmodern Body": "Passing implies pretense and lying, not a new ontological reading of the body's sex. Postmodernism with its notion of the body as invention of discursivity makes plausible the 'transsexualizing' of the body, a possibility dependent on the adoption of new criteria and alternative readings of the body's sex" (1998, p. 106). Sandy Stone, in a profoundly different vein, discusses the need for transsexuals to resist the impulse to pass so that they may form political constituencies as *transsexuals* in "The *Empire* Strikes Back: A Posttranssexual Manifesto." In this endnote, I do not wish to suggest that femme passing and transsexual passing are the same phenomena. Rather, I hope that their contiguities, affiliations, coalitions, and differences might be drawn out with increased specificity. No doubt, their connection, even if it is only a semantic one denoted by the term "passing" itself, is not written about enough by feminist and queer theorists.

3. Throughout the course of this paper, I make a distinction between "transgender" and "transsexual" subjects. For my purposes, transsexual subjects are those subjects who specifically undergo anatomical shifts to assuage their feelings of being wrongly sexed. Many may see my distinction between transgender and transsexual subjects as

arbitrary if not entirely false, but the reasons for such a distinction will become more evident later.

4. That is, mapping out our sexed identities (male, female, etc.), gender identifications (man, woman, trans, etc.), sexual identifications (heterosexual, homosexual, bisexual, etc.), sex/gender of object choices (male, female, man, woman, trans, etc.), and desires (heteroerotic, homoerotic, queer, etc.) in the most *heteronormatively* aligned fashion possible. Judith Butler very astutely outlines this alignment in her first chapter of *Gender Trouble: Feminism and the Subversion of Identity*, "Subjects of Sex/Gender/Desire," pp. 3–44.

5. This idea that sex is just as fluid as gender if not more so has come from a series of discussions I have had with Elizabeth Grosz. The fluidity of sex that I speak of in this essay is her idea. My particular exposition of sex's fluidity, though, is my responsibility and mine alone.

6. This is not to suggest that lesbian, queer, and feminist scholars do the same theoretical and political work. Their projects are oftentimes very divergent and are aligned with profoundly different theoretical and political traditions, although I do think that a significant shared fissure for all of them is the exploration of transsexual subjects and their subjectivity.

7. See Detloff, Madelyn (2006). Gender Please, Without the Gender Police: Rethinking Pain in Archetypal Narratives of Butch, Transgender, and FTM Masculinity (this volume).

8. Judith Halberstam discusses the phenomenon of the masculine continuum in two pieces. The first is (1998). "Transgender Butch: Butch/FTM Border Wars and the Masculine Continuum." *GLQ. The Transgender Issue*. 4(2). Durham: Duke University Press, pp. 287-309. She republishes another version of this essay in with the same title in (1998). *Female Masculinity*. Durham: Duke University Press, 1998, pp. 141-174. In my commentary with Angela Pattatucci-Aragón, she very astutely suggests that the queer overemphasis on notions like "continuum" unfairly erases the cultural intelligibility of femme men who are not gay-identified. Their cultural intelligibility results from either entering into gay communities where their femmeness is acknowledged to the extent that it can be sexualized *or* their femmeness is acknowledged by surgically entering into a biologically female body. While inclusive of female butchness, continuums in queer theory have tended to ignore male femininity.

9. By using the term "campy" in this sentence I do not mean to downplay the serious struggles entailed in transgender embodiment. Campiness can entail a great deal of pain and suffering too. See specifically Leslie Feinberg's *Stone Butch Blues: A Novel*.

10. With this claim, I do not mean to suggest that the term "transgender" should be set to *only* represent the lived pains and struggles of those wishing to fall within its rubrics. Nor do I intend to suggest that the term "owns" the pain in gender transitioning or in queer practices. Rather, I am generalizing to make a point about how the term "transgender," as of late, seems to be continually associated with pleasurable valences. For a more sustained discussion and analysis of Butch/FTM pain, please see the work of Madelyn Detloff in this volume, "Gender Please, Without the Gender Police: Rethinking Pain in Archetypal Narratives of Butch, Transgender, and FTM Masculinity."

11. For a concise version of this history see Cressida J. Heyes' essay "Feminist Solidarity After Queer Theory: The Case of Transgender."

12. Halberstam has since revised her seminal formulation "There are no transsexuals. We are all transsexuals" (1994, p. 212). The infamous quote originally appeared in "F2M: The Making of Female Masculinity" in the anthology *The Lesbian Postmodern*

edited by Laura Doan. Her well-received corrective statement appears in "Transgender Butch": "There are transsexuals, and we are not all transsexuals; gender is not fluid, and gender variance is not the same wherever we find it. Specificity is all. As gender-queer practices and forms continue to emerge, presumably the definitions of *gay, lesbian, transsexual,* and *transgender* will not remain static, and we will produce new terms to delineate what the current terms cannot" (1998, p. 306).

13. Many scholars, especially ones articulating the fragmented and transitory conditions of postmodernity, have utilized transsexuality for its ability to be paradigmatic of the postmodern condition. Rita Felski documents this phenomenon in her essay "Fin de siecle, Fin de sexe: Transsexuality, Postmodernism, and the Death of History." She argues that famous postmodern theorists like Derrida, Deleuze, and Baudrillard use transsexuality as a trope for the end of history, an atemporal condition touted by postmoderns. Her essay eloquently asks, "How do our cultural imaginings of historical time relate to changing perceptions of the meaning and nature of gender difference?" (1996, p. 337). "Fin de siecle, fin de sexe" is an epigram coined by French artist Jean Lorrain to denote the figurative affinity of history's end with sex's transitional character. Felski describes the utility of this phenomenon at length, "Thus the destabilization of the male/female divide is seen to bring with it a waning of temporality, teleology, and grand narrative; the end of sex echoes and affirms the end of history, defined as the pathological legacy and symptom of the trajectory of Western modernity" (1996, p. 338). Jean Baudrillard's works *The Transparency of Evil* and *Cool Memories* are perhaps the most famous examples of this phenomenon. Paradigms of gender and sexuality emerge for him to allegorize the epidemic of signification that *is* the (anti)ontological condition of postmodernity. Other texts that document the "free-floating aestheticism" of sex and gender signifiers are Arthur and Marilouise Kroker's anthologies *Body Invaders* and *The Last Sex*. Despite profound acclaim for her essay, Felski has been criticized by transsexual theorists like Jay Prosser in *Second Skins: The Body Narratives of Transsexuality* for her failure to adequately specify transsexual and transgender subjects, as evidenced in this quote. "For example, in *The Transparency of Evil*, Jean Baudrillard writes, 'the sexual body has now been assigned an artificial fate. This fate is transsexuality–transsexual not in any anatomical sense but rather in the more general sense of transvestitism, of playing with the commutability of the signs of sex . . . we are all transsexuals.' Here transsexuality, or perhaps more accurately, transgenderism, serves as an overarching metaphor to describe the dissolution of once stable polarities of male and female, the transfiguration of sexual nature into the artifice of those who play with the sartorial, morphological, or gestural signs of sex" (1996, p. 337).

14. My use of the term "queer" in this sentence is not to suggest that all lesbians identify as queer or all queers identify as "lesbians." Rather, in this sentence I wish to capture the considerable variation of lesbian bodies and bodily practices by using "queer."

15. But like most archaeological rupturings, these texts were taken up in deeply arbitrary, unpredictable, and, at times, problematic ways. Certainly scholars have done much in the way of criticizing the larger epistemological presuppositions of each text as well as problematizing their more specific questionable maneuvers, but critique most assuredly fails to prevent others from misappropriating both well-thought-out and not so well-thought-out arguments.

16. The bulk of this theorizing occurs in her second chapter, "Prohibition, Psychoanalysis, and the Production of the Heterosexual Matrix." Briefly, she begins with Lacan's theory of language, where, via paternal law, men are said *to have* the phallus

while women are said *to be* the phallus. Butler reads Lacan as arguing that these onto-logical specifications of having and being the phallus are never fully actualized by men and women, respectively; having and being are ultimately "comedic failures that are nevertheless compelled to articulate and enact these repeated impossibilities" (1990, p. 46). Because women fail to ever fully become the phallus, they are always an ap-proximation of sorts–they merely "appear" to be the phallus. Lacan, terms this "ap-pearing as being the phallus" masquerade. The function of masquerade, according to Lacan is to mask or '[dominate] the identification through which refusals of love are re-solved' (Lacan, quoted by Butler, 1990, p. 48), the mechanism of Freud's melancholia. Butler turns to Freud's notion of melancholia and reads it as a gender ontology for two reasons. First, because of melancholia's reliance on the Oedipal drama, and the Oedi-pal drama's basis in an externally enforced prohibition against a son's sexual desire for his mother and father, Butler is able to argue that the process of masquerade (with its reliance on melancholia) is not attempting to access a signification prior to culture. Whereas Lacan awards the realm of the Symbolic special access to the 'real' prior to the realm's emergence in culture, Butler argues that masquerade's dependence on mel-ancholia and melancholia's dependence on externalized prohibitions requires a reconceptualization of masquerade. The continual play of appearances that make up the "internalized" dramas of masquerade can no longer be conceived of as based in an ability to approximate a pre-cultural ideal of the phallus. Rather, according to melan-cholia, cultural taboos on incest and homosexuality produce the dramas of the psyche. Therefore, these dramas are not framed by a pre-cultural, pre-discursive ideal. The sec-ond reason Butler returns to Freud's notion of melancholia is to argue that the mecha-nism of melancholia explains how one's truth of sex comes to be located on the surface of the body. Whereas the notion of masquerade temporalizes the play of the appearance of gender, melancholia theorizes the production of an incorporated space, where one's lost love (i.e., father and/or mother) and lost modality of desire (i.e., heterosexual or *homosexual* as Butler theorizes) becomes internalized. This incorporated space, ac-cording to Butler, is located on the surface of the body. Melancholia, the never-ending process of identity formation through incorporating the history of one's lost love ob-jects, explains for Butler the incessant acquisition of one's gender identity.

17. For the purposes of my investigation of "transgender" in feminist scholarship, two things about Butler's work are important to note. The first has just been discussed at length; that is, the overt privileging of psychoanalysis to discuss gender's unnatural-ness and transitional character (*but not a transitional character that is easily or voli-tionally entered into*). The second is Butler's wedding of false sexed ontology to a matrix of compulsory heterosexuality. For Butler gender and (sex)uality are *always al-ready* inter-articulated in highly complicated and interdependent ways.

18. Biddy Martin superbly accounts for this phenomenon in her aforementioned "Sexualities Without Genders and Other Queer Utopias." She begins her comprehen-sive article, "For a long time I have been concerned about a tendency among some les-bian, bisexual, and gay theorists and activists to construct 'queerness' as a vanguard position that announces its newness and advance over and against an apparently super-seded and now anachronistic feminism with its emphasis on gender. [. . .] But I am worried about the occasions when antifoundationalist celebrations of queerness rely on their own projections of fixity, constraint, or subjection onto a fixed ground, often onto feminism or the female body, in relation to which queer sexualities become figural, performative, playful, and fun" (1994, p. 104).

19. This quote is taken by C. Jacob Hale's "Consuming the Living, Dis(re)membering the Dead: In the Butch/FTM Borderlands," 1998, p. 311.

20. For a review of several of these books see Bernice L. Hausman's "Recent Transgender Theory."

21. In my commentary with Angela Pattatucci-Aragón, she insightfully expands my argument here to suggest that not only is feminism's uptake of the transgender subject position about appeasing the post-identity impulse of queer theory, but it also explores new possibilities of being and doing womanhood whereby patriarchal power might be increasingly diminished and undermined. Transgender undoubtedly proliferates the times and spaces of patriarchal resistance, but it will by no means end patriarchy by overturning the given conceptual status of gender's binarity. Indeed, the struggle against patriarchy is much more multidimensional than the scope of a conceptual or linguistic shift.

22. This idea of reontologizing sex comes from a conversation I had with Elizabeth Grosz. What this reontologization entails for transsexuals, though, is my idea alone, and I take full responsibility for its potential problems.

23. Bernice L. Hausman's *Changing Sex: Transsexualism, Technology, and the Idea of Gender* has been criticized on several fronts for positioning transsexuals as cultural dupes of the medical establishment. See most particularly the introduction to Jay Prosser's *Second Skins: The Body Narratives of Transsexuality*, "On Transitions–Changing Bodies, Changing Narratives" and Judith Halberstam's "Transgender Butch: Butch/FTM Border Wars and the Masculine Continuum."

24. While my emphasis is primarily on transsexual subjects, poststructuralist feminists, by failing to engage the biological sciences, also have difficulty addressing the issue of a woman who reinvents or enters more fully into her femininity via plastic surgery. A whole range of theoretical and political issues and questions arise through biology's capacity to shift the body's fleshy contours that poststructuralist feminists have considerable difficulty addressing. Transsexuals are obviously not the only sexed and gendered subjects who feel uncomfortable with their sexed assignment at birth. Even anatomical females who remain female look to biology to augment or improve their femininity in culturally normative ways.

25. My skeptical reader continues to ask, "Why biology for transsexuals?" Because biology is exactly the locale where transsexuals garner agential specificity from other marginally gendered bodies. Other theorists of transsexual subjectivity reconsider phenomenology and its capacity to theorize notions of sexed (dis)embodiment and interiority that poststructuralist accounts of gender gloss over. Phenomenology seems to be the best scholarly venue from which to study *feelings* of sex *dis* and *re* embodiment for transsexuals, before, during, and after surgery that exceed discursive prediction. In "Phenomenology as Method in Trans Studies," Henry S. Rubin argues that "whereas discursive analysis hopes to penetrate essences and demonstrate the fiction of their fixed and naturalized character, phenomenology takes it as a matter of fact that essences are always already constituted in relationship to embodied subjectivity, hence they are natural and malleable" (1998, p. 267). Jay Prosser vies for phenomenology as well in theorizing transsexual subjectivity in the second chapter of his *Second Skins: The Body Narratives of Transsexuality*, "A Skin of One's Own: Toward a Theory of Transsexual Embodiment," pages 61-98. Two other important pieces bring up the problematic of embodiment for theorizing transsexual agency: Bernice L. Hausman's "Virtual Sex, Real Gender: Body and Identity in Transgender Discourse" and Patricia Elliot's and Katrina Roen's "Transgenderism and the Question of Embodiment: Prom-

ising Queer Politics?" As critics of discursive analysis have noted, discourse, especially language, binds the expression of affect, or the feeling part of emotion. See
particularly Eve Kosofsky Sedgwick's *Touching Feeling: Affect, Pedagogy, and
Performativity*. Discursive analysis cannot get at transsexuals' plea that they feel in the
wrong body or wrongly embodied, and phenomenology understands bodily ontology
in such a way that is not overly determinative or necessarily sex-essentialist. Specifically, theorists of transsexuality have been most interested in phantom limbs (i.e., in
the case of an FTM, imagining a penis that does not exist) or anosognosia (i.e., in the
case of an MTF, having a penis that one does not want or does not feel as one's own).
How can feminist and queer theorists unproblematically valorize sexual pleasure when
many subjects do not *feel* right with the anatomical apparatuses that such pleasure ostensibly coheres around? This is not to suggest that feminist and queer scholars not theorize pleasure. Rather, it is to note that the conditions of pleasure are themselves highly
contingent and in need of more extensive consideration. Brushed under the
poststructuralist feminist epistemological and methodological rug as of late, phenomenology, in its ability to take seriously embodiment and interiority, certainly deserves
reconsideration, but it does not get at transsexual specificity in quite the same way that
biological sciences do. Feminist and queer theorists may use phenomenology to theorize a whole set of gender-queer bodies that claim sexed disembodiment, but biology
specifically captures transsexual agency in fleshy transitions like no other disciplinary
matrix.

REFERENCES

Baudrillard J (1990). *Cool Memories*. London: Verso.
Baudrillard J (1993). *The Transparency of Evil: Essays on Extreme Phenomena*. New
    York: Verso.
Bornstein K (1994). *Gender Outlaw: On Men, Women, and the Rest of Us*. New York:
    Routledge.
Butler J (1990). *Gender Trouble: Feminism and the Subversion of Identity*. New York:
    Routledge.
Cameron L (1996). *Body Alchemy: Transsexual Portraits*. Pittsburgh: Cleis Press.
Elliot P and Roen K (1998). Transgenderism and the Question of Embodiment: Prom-
    ising Queer Politics? In: Stryker S (Ed.), *GLQ. The Transgender Issue* 4(2),
    231–261.
Feinberg L (1993). *Stone Butch Blues: A Novel*. Los Angeles: Alyson Books.
Felski R (1996). "Fin de siecle, Fin de sexe: Transsexuality, Postmodernism, and the
    Death of History." *New Literary History* 27, 337-349.
Garber M (1992). *Vested Interests: Cross-Dressing and Cultural Anxiety*. New York:
    Routledge.
Halberstam J (1994). F2M: The Making of Female Masculinity. In: Doan L (Ed.), *The
    Lesbian Postmodern*. New York: Columbia University Press.
Halberstam J (1998). *Female Masculinity*. Durham: Duke University Press.
Halberstam J (1998). Transgender Butch: Butch/FTM Border Wars and the Masculine
    Continuum. In: Stryker S (Ed.), *GLQ. The Transgender Issue* 4(2), 287-309.

Halberstam J and Hale CJ (1998). Butch/FTM Border Wars: A Note on Collaboration. In: Stryker S (Ed.), *GLQ. The Transgender Issue 4*(2), 283-285.

Hale, CJ (1998). Consuming the Living, Dis(re)membering the Dead: In the Butch/FTM Borderlands. In: Stryker S (Ed.), *GLQ. The Transgender Issue 4*(2), 311-348.

Hausman BL (1995). *Changing Sex: Transsexualism, Technology, and the Idea of Gender*. Durham: Duke University Press.

Hausman BL (1999). Virtual Sex, Real Gender: Body and Identity in Transgender Discourse. In: O'Farrell MA Vallone L (Eds.), *Virtual Gender: Fantasies of Subjectivity and Embodiment*. Ann Arbor: The University of Michigan Press, pp. 190-216.

Hausman BL (2001). Recent Transgender Theory. *Feminist Studies 27*(2), 465-490.

Heyes JC (2003). Feminist Solidarity After Queer Theory: The Case of Transgender. *Signs 28*(4), 1093-1120.

Kroker A and Kroker M (Eds.) (1987). *Body Invaders: Panic Sex In America*. New York: St. Martin's Press.

Kroker A and Kroker M (Eds.) (1993). *The Last Sex: Feminism and Outlaw Bodies*. New York: St. Martin's Press.

Martin B (1994). Sexualities without Genders and Other Queer Utopias. *Diacritics. Special Issue on Critical Crossings 24*, 104-121.

Prosser J (1998). *Second Skins: The Body Narratives of Transsexuality*. New York: Columbia University Press.

Raymond J (1979). *Transsexual Empire: The Making of the She-Male*. Boston: Beacon Press.

Rubin HS (1998). Phenomenology as Method in Trans Studies. In: Stryker S (Ed.), *GLQ. The Transgender Issue 4*(2), 263-281.

Sedgwick EK (1990). *Epistemology of the Closet*. Berkeley: University of California Press.

Sedgwick E (2003). *Touching Feeling: Affect, Pedagogy, Performativity*. Durham: Duke University Press.

Stone S (1991). The *Empire* Strikes Back: A Posttranssexual Manifesto. In: Eptein J and Straub K (Eds.), *Body Guards: The Cultural Politics of Gender Ambiguity*. New York: Routledge, pp. 280-304.

Stryker S (1998). The Transgender Issue: An Introduction. In: Stryker S (Ed.), *GLQ. The Transgender Issue. 4*(2), 145-158.

Zita JN (1998) *Body Talk: Philosophical Reflections on Sex and Gender*. New York: Columbia University Press.

The Invisible Body of Queer Youth: Identity and Health in the Margins of Lesbian and Trans Communities

Dorinda L. Welle
Sebastian S. Fuller
Daniel Mauk
Michael C. Clatts

SUMMARY. How does complexity in gender and sexual identity construction and partnering practices generate unique vulnerabilities for queer-identified youth? We present two case studies from an ongoing ethnographic study of LGBTQ youth development: "Samantha," a queer-

---

Dorinda L. Welle, PhD, is Director of the Youth and Community Development Research Core in the Institute for International Research on Youth at Risk at the National Development and Research Institutes, Inc.

Sebastian S. Fuller, MA, is a Research Associate in the Institute for International Research on Youth at Risk at the National Development and Research Institutes, Inc.

Daniel Mauk, MPH, is a Research Associate in the Institute for International Research on Youth at Risk at the National Development and Research Institutes, Inc.

Michael C. Clatts, PhD, is Director for the Institute for International Research on Youth at Risk at the National Development and Research Institutes, Inc.

Address correspondence to: Dorinda Welle, International Research on Youth at Risk, National Development and Research Institutes, Inc., 71 West 23rd Street, 8th Floor, New York, NY 10010 (E-mail: welle@ndri.org).

[Haworth co-indexing entry note]: "The Invisible Body of Queer Youth: Identity and Health in the Margins of Lesbian and Trans Communities." Welle, Dorinda L. et al. Co-published simultaneously in *Journal of Lesbian Studies* (Harrington Park Press, an imprint of The Haworth Press, Inc.) Vol. 10, No. 1/2, 2006, pp. 43-71; and: *Challenging Lesbian Norms: Intersex, Transgender, Intersectional, and Queer Perspectives* (ed: Angela Pattatucci Aragón) Harrington Park Press, an imprint of The Haworth Press, Inc., 2006, pp. 43-71. Single or multiple copies of this article are available for a fee from The Haworth Document Delivery Service [1-800-HAWORTH, 9:00 a.m. - 5:00 p.m. (EST). E-mail address: docdelivery@haworthpress.com].

Available online at http://www.haworthpress.com/web/JLS
doi:10.1300/J155v10n01_03

identified woman partnered with a transgender man, and "Reid," a queer-identified transgender man who has declined medical gender transitioning and who partners with lesbians and gay men. We consider the implications of these youths' locations on the margins of both lesbian and transgender communities and the challenges in providing health care and support services for queer-identified youth. *[Article copies available for a fee from The Haworth Document Delivery Service: 1-800-HAWORTH. E-mail address: <docdelivery@haworthpress.com> Website: <http://www.HaworthPress.com> © 2006 by The Haworth Press, Inc. All rights reserved.]*

KEYWORDS. Queer youth, transgender youth, transgender identity, transgender health, language practices, marginalization, HIV risk

The recent and historic special issue on Lesbian, Gay, Bisexual, and Transgender (LGBT) Health, published by the *American Journal of Public Health* in June 2001, not only aimed to "add transgender health" to an existing LGB health agenda, but presented a concerted approach to LGBT health, from medical care and public health perspectives. In particular, Lombardi (2001) emphasized the need to "allow for complexities" in providing transgender health care: a guideline that should well be extended to any approach to LGBT health. While LGBT health services provide a unique and critical "site" for engagement of lesbian, gay, bisexual, transgender, and queer-identified individuals with a range of locations in LGBT "communities" and roles in the system of health services, here we explore the specific invisibility of "queer-identified" youth and consider their unique needs for support in accessing primary and LGBT-specific health care and mental health services.

Reporting on an ongoing study of LGBTQ youth in New York City, this paper presents two ethnographic case studies: one of a queer-identified woman (Samantha) and another of a "transgender queer"-identified young man (Reid), both of whom are variously located on the margins of lesbian *and* transgender communities. Situated within diverse economic and educational backgrounds, their narratives illustrate the simultaneous centrality and marginality of "the body" in presentations of "queer" gender styles, sexual attractions, and identity politics. In addition, what could be called "queer" discourses, whether originating in the academy or on the street, juxtapose gender and sexuality references in such a way as to disrupt a unified notion of "the body," generating new challenges for accessing and utilizing health care services. Youth in gender transitions and youth in fluid identifications speak of "queer" experience while still referencing "normal" experience, further complicating their perspectives on how and where to seek health care. Subsequently, the politics of basic and specialized needs for health care and social support services that is a central concern and historic achievement in lesbian communities and emerging transgender communities may fall outside the radar of queer-identified youth on the margins of both.

In public health research, an understanding of "queer" identification remains elusive. Studies that are inclusive of LGB *and* T individuals are few, although some studies of LGB youth classify all participants as queer youth, using the term "queer" as a substitute for LGB, generating another linguistic exclusion of transgender youth. Intersex youth have been thoroughly invisible in public health research; two years into our current study, one transgender woman identified herself as intersex. Queer identifications may be differently distributed across racial or ethnic groups. The national Black Pride Survey (Battle et al. 2000) found that only 1% of 2,500 Black LGBT people attending Black Gay Pride celebrations self-identified as queer. "Queer" may also be considered an insider term operant "within" LGBT communities: what Connell (1995) has called "a dissident politics from within," and hence difficult to capture through survey data.

Darryl Hill (2000) has noted that the "either/or logic" of gender and sex yield "either/or" identifications, "both/and" identifications, and "neither/nor" identifications. With respect to our two case studies, Reid described "queer" as a combination of "old school plus new school," drawing from the various ways that lesbians, gay men, and transgender men have challenged gender and sexuality stereotypes. Samantha considered "queer" as a commitment to fluid sexuality and gender identifications, an artifact of combinations and rejections.

Samantha currently dates a transgender man and Reid identifies as a transgender man. Both of these youth identify as "queer," and both have sexually partnered with women and transgender men. Reid has also reported partnering with gay men. Initially, we viewed these two case studies as important illustrations of how self-identifications and social affiliations with transgender men placed LGBTQ youth "on the margins" of both lesbian and transgender communities. However, queer self-identifications also significantly inform marginality. Illustrating how Reid and Samantha constructed their own understandings of "lesbians," "fagged-out women," "tranny boys," and "trans-men," we examine how each combines gender imagery and imagery of lesbians and gay men in the construction of clothing styles, gender styles, attractions, and selves. We consider the implications of these narratives for the provision of social supports and health services for queer-identified youth and transgender-identified youth from diverse economic and educational backgrounds. Throughout, we aim to create an analytic space that can foster an appreciation of the contradictions, challenges, and processes involved in the pursuit of a "different" (or "very different") self.

QUEER-IDENTIFIED AND TRANSMALE-IDENTIFIED YOUTH ON THE MARGINS OF HEALTH SERVICES

Gender-queer-identified and Transmale-identified youth often struggle for language to describe their own or their transgender peers' or partners' understandings of themselves. It is important to understand this struggle for descriptive language in historical context. On the one hand, feminist health advocacy historically has been intertwined with projects to theorize gender and sexuality (e.g., Rubin 1975; Gordon 1976; Davis 1981; Joffe 1987). However, the significance of earlier achievements in terms of theorizing, developing gender-sensitive health care approaches, and advocating for reproductive rights is often overlooked in critiques of the largely "second wave" orientation of 1970-1980s feminist health advocacy and post-structuralist rejections of "women" as a unified gender category. Subsequently, in an era when "care of the body" has been theorized as a modern form of discipline (Foucault 1977), feminist health may not represent the kind of contemporary "radical" politics as it did for previous cohorts.

At a time when feminist health movements have been stereotyped as "old school," the "new school" movements, including the emerging field of transgender health (Bockting et al. 1998; Futterweit 1998; Israel & Tarver 1997; Karasic 2000; Lombardi 2001; Oriel 2000), are still in formation. Before 1991, there was virtually no literature available on Female-to-Male transsexuals (FTMs) or transmen (Cromwell 1999). After 1995, publication of books specific to transmale issues has grown significantly, affording increased visibility in less than a ten-year period. Jason Cromwell's (1999) *Transmen & FTM's* and Henry Rubin's (2003) *Self-Made Men* seek to convey transmale experience through qualitative data and sociological theory. Autobiographical and biographical works include *A Self-Made Man: The Diary of a Man Born in a Woman's Body* (Hewitt & Warren 1997) and Loren Cameron's (1996) photo-documentary-style *Body Alchemy*. *The Phallus Palace* (Kotula 2002) represents an amalgam of these two genres, using short theoretical essays, autobiographical writing and portraiture, as well as interviews with surgical professionals and photo documentaries of various surgical techniques. It is significant that within all of these publications there is an intrinsic medical conversation and exploration of available health services. Hewitt describes the medical procedures he undergoes, Cameron photographs the surgically and hormonally altered transmale body, and Rubin (2003, 93-113) speaks of body disassociation and medical modification.

The Internet is a growing source of information about specific surgical and medical masculinizing procedures.[1] College-educated transgender men in our study were more likely to augment Internet-based information about gender transitioning with the advice of clinicians, and articulated their own critical perspectives on trans-health, including critiques of The Benjamin Standards of Care.[2] In contrast, non-college-educated transgender men like Reid were more likely to access Internet resources in a fragmented and isolated way, seeking medical information but avoiding online discussions and listservs, and expressing reluctance to access transgender health services.

THE 2GROWN PROJECT:
LGBTQ YOUTH IN DEVELOPMENTAL CONTEXT

The 2Grown Project is a 4-year longitudinal ethnographic[3] study of developmental complexity[4] among street-involved lesbian, gay, bisexual, transgender, queer and questioning youth. Recruiting "questioning" youth enabled us to include those who "rejected" LGBT identities or were not sure of their sexual or gender orientations, but were partnering with same-gender partners or were actively presenting as other than their birth-assigned gender. Employing life history interviewing, participant-observation, and ethnographic follow-up interviewing focused on sexual and drug-use-related risk, protective behaviors, and participants' life history themes, the study recruited youth through street-based public venues, including internet cafes, bars and clubs, after-school hangouts, and "The Piers"[5] in Greenwich Village. Consequently, the cohort of 45 LGBTQ youth (ages 17-22 at baseline) is particularly diverse in terms of economic and educational locations, including homeless youth, youth living with parents, youth living with partners, and youth living in subsidized or emergency housing. Participants include high school students, recent dropouts, college and graduate students. They obtain income from diverse sources as minimum wage employees, as college graduates in their first "real" jobs, as recipients of parental support, and as drug sellers and street-based commercial sex workers.

Significantly, nearly all of the queer-identified youth in the study were individuals of birth-assigned female gender. Identifications included gender queer, queer women, and queer transmen. While several queer-identified youth had previously engaged in reproductive rights activism, lesbian health activism, and/or transgender health activism, their life history narratives were strikingly silent about their own "personal" health concerns. General health information typically required formal prompting by the ethnographers.

Sexual and gender identities are constructed through diverse sets of literacy and sexuality practices that employ both ideological and "naturalized" discourses that variously illuminate and mask the dimension of health (Welle, Clatts & Barnard, 2005). Important differences emerge in how LGBTQ youth in the study define, discuss, and engage queer and transgender identities. Specifically, two types of discursive frames are evident in LGBT youths' narratives: namely, what Chiseri-Strater (1991) has called "academic literacies" which are taught and elaborated in and beyond the university setting, and what we call "Pier/Peer literacies"–local, street-based discursive practices that emerge in the local setting of the Christopher Street/West Side Highway Piers, where a large number of LGBT youth of color and homeless youth "hang out." Both of these literacies challenge the limits of "lavender languages" or "gay language" (Leap & Beollstorff 2003) in the reinvention of identities (Livia & Hall 1997; Bucholtz et al. 1999), specifically by shifting the discursive project from one of *identity formation* to the broader project of *constructing a self*. As the case studies below illustrate, discursive practices are not only employed in the interests of theory, politics, or community, but also (for some) to describe the construction of a "queer" self that feels at once "different" and at once a "normal," "natural" self.

Those gender-queer-identified and transgender-identified youth who attended private colleges had all participated in critical gender studies and queer studies courses and referenced academic discourses to frame LGBT and queer identities. Queer-identified and transgender-identified college students in our sample selectively used elements of postmodern theories to fashion a working theory of the self: a project that most postmodernists would openly reject. As a result, the narratives of the queer-identified and transgender-identified college students struggle against the fundamental contradiction of utilizing post-modernist language to describe the fairly modernist construct of "the self." Many used the language of deconstruction in a process of self-construction, but with little awareness of how their own subjectivity was "torn in two directions" in the process (Welle, Clatts & Barnard 2005). In addition, many of the queer-identified and transgender-identified college students sought to reject and go beyond what they experienced as the limits of institutionalized gay and lesbian cultures in the university–what one youth from a co-ed university called "the gay supermarket" and another from a women's college called "lesbionic space."

Not all of the queer-identified and transgender-identified youth in the study attended college or had access to academic discourses about gender and sexuality. Several in fact had dropped out of high school, left home, or were kicked out of their homes as young teenagers, and were recruited into the study while homeless. Queer-identified and transgender-identified homeless youth reported a range of educational aspirations, including getting GEDs, finishing high school, and/or attending college. Their narratives variously reference terminology learned from their peers on the street in Greenwich Village, "program language" in use at the nearby Lesbian, Gay, Bisexual, and Transgender Community Center's youth services program, and discursive resources drawn from television, music, and to a lesser extent the Internet. In the case studies below, we have chosen not to edit the youths' language, but rather to represent their struggles to communicate multiple locations, voices, and perspectives.

SAMANTHA: QUEER POLITICS AND THE AFTER-SCHOOL SNACK

Samantha is a 22-year-old Jewish woman who grew up in an upper-middle-class family. She arrives for the baseline interview boasting a pair of "charcoal gray" jeans, a pair of vintage shoes, and an "awesome white sweatshirt," aiming for a style that's "punk rock but a little girly . . . sort of like punk rock plus *Sex and the City*, put it all together." Recollecting her early experiences of being attracted to other girls in elementary school and heading the Gay-Straight Alliance in high school, she can easily

> still feel like I'm twelve years old and I'm the one who's gay, like, this really, like, shameful thing, you know? . . . back in that "I'm embarrassed," like, mindset, you know.

Her parents "overlook (her) gayness" and also remain unaware that Samantha was hospitalized for an eating disorder as a young teenager. After the hospitalization, she became vegan, "which is also this, like, kind of restrictive, like, (behavior). I feel better about myself if I can, like, limit my food intake in this way, or be in control." Although she "still, totally, like, (has) these little episodes" of disordered eating, she is not currently receiving any therapy or support services. She attributes her eating disorder to the fact that her body developed "early," with "people treating (her) in a really sexual way." Samantha recently took a course in which she reflected further on eating disorders.

> I took this really awesome class last year, with, like, my favorite professor, about psychoanalysis and whiteness, and about, like, and so now I think a lot about (the eating disorder). We study all these things about, like, whiteness and, like, neuroses, and, like, craziness and things that happen to, like, the body and personality because of this kind of, like, culture of whiteness that we're all, you know, like, we were talking about how that works, and how, like, eating disorders are, like, a part of that, and, like, control and neuroses, and whatever. It's like, a longer theory, but anyway. I feel that totally made sense.

Similar to many of the other queer-identified college students in the study, Samantha approaches college with competing if not contradictory interests: on the one hand, seeking answers to "the self" through cultural theory while simultaneously decentering the self through these same critical discourses. While theory helps her "make sense" of her eating behavior, friends played a critical role in helping her emerge from a two-week period of self-starvation after a recent breakup with a boyfriend.

A self-described "girly" woman dating a "tranny boy," Samantha describes the development of her queer politics through a rejection of liberal feminism while attending a prestigious women's college.

> I guess when I was starting to get involved in, like, queer activism and, like, feminist stuff, it was in this very, it was, like, among a particular group of people who had, like, kind of mainstreamy sort of '70s feminist, like, ideas. . . . Like, the kind of feminism I would teach was all, like, equality and equal rights, and that, sometimes it doesn't seem to, like, work all the time, you know? Like, 'cause there's always equality for some people and, like, whatever. It's

about, because I think really interesting, like, playing around with power in a more complex way, you know? It was, like, not just about some really simplistic, like, 'We should all be the same! We should be asking for the same thing!' 'Cause it's just not true, you know?

Making a quick pass through essentialist or separatist feminist politics, Samantha quickly and "thankfully" moved into reproductive rights activism and then, in college, became involved in trans health and queer rights activism.

Like, I've been doing a lot of activist work since I was, like, fifteen. And I started, like, this really, like, traditional hardcore feminist way, like, men are evil, feminists are great. That only lasted, like, a year, thankfully (laughs). We didn't get influenced. And then, we did a lot of like, reproductive rights activism, and, like, kind of health stuff. And then, when I was at school, I did a lot of trans health, and like, queer rights stuff.

While in college and volunteering at a women's rights organization, Samantha helped organize a talk by a well-known writer and S/M advocate.

You know, the lady who writes the, whatever. And she came to speak, and then her girlfriend at the time read, like, was wearing, like, leather pants and, like, leather top, and totally, like, all decked out in this, like, definitely the most, like, queer, like, out-radical-looking person I'd ever seen, you know?

Intrigued with the "out-radical-looking" style of S/M leatherwear, Samantha and a girlfriend attended their first S/M party when they "totally weren't old enough."

It was at this dungeon in Brooklyn, and it was like, a women's S/M party. So we were, like, 'Oh, let's go.' So we went, and we totally weren't old enough to go. And then, like, and we went. It was just, like, really, like, shocking and interesting and, like, wow. You know? Like, I had definitely never, ever seen before, and it just seemed to be, like, playing with power dynamics in a really different way.

Samantha describes the urban space of New York City and the "shocking and interesting" space of the S/M dungeon as "out" spaces that could support an experience of liberation.

> It was so fucking liberating to be in New York City for the week-end, and like, to be in a room of people who were, like, playing around, or doing things, like, totally, like, in "out" space, you know? It was just very, like, liberating, you know?

Samantha and her boyfriend Ethan engage in sexual role playing-"even if it's just in our apartment"–that relies on language of "tops" and "bottoms" often used to describe gay men's sexual and/or social roles. However, while Samantha invests the terms "topping" and "bottoming" with meanings about dominance and submission, her use of these terms also leaves ambiguous the sexual practices in which each partner engages.

> So it's like, with Ethan, we whatever, we play around and I, like, I bottom people, like, it's not something that I wouldn't do, but, like, with Ethan, we have this, like, kind of mommy-boy relationship, and, like, he has a collar, and he's the cutest little thing ever, and I really like, like, I really like the whole, I like the whole mommy's boy thing so much . . . and having this kind of, like, nurturing whatever thing happening, it was okay. So with him, it's really, really nice, because, like whatever. I do really cheesy, exciting things with him. I can make him after-school snacks, and, like, it makes me really happy, and I feel really nice, and he feels all, like, taken care of. And whatever, and we take care of each other in our own, like, normal relationship. And it's really nice. . . . And there's something about the, like, tabooness of it that I'd be able to, like, indulge it, and like, when he says to me, like, "Mommy, can I have la la la la la," or whatever, that makes my heart, like, jump and feel really excited, you know.

Samantha's descriptions of "bottoming" for Ethan through "indulging" in "nurturing" behavior also mask a life history context that further situates the meanings of these role assignments. In her own life history context, occupying the role of food provider and being the person in control of satisfying hunger or withholding food represents for Samantha an exciting, "happy," "really nice" experience.

In addition, her assumption of the role of food provider has contemporary significance in self-development. Preparing to soon finish college, she herself may be anticipating her own "after-school" rewards. Unsure whether these will be readily forthcoming in an uncertain economic climate, Samantha nonetheless "plays" with the possibility of being able to both "tame" her desires and satisfy them (also see Welle et al. in press). Whereas Samantha felt that liberal feminism insisted that "everyone be asking for the same thing," she and Ethan have created a "space" in which the request for an after-school snack can give voice to diverse powerful processes: an assertion of a child's "right" to control and satisfy her bodily needs, the assumption (and overthrow) of the potentially terrible powers of the withholding Mommy for whom "nurturing" is viewed as "taboo," and the opportunity to re-script the relationship of eating behaviors and sexuality.[6]

Considering what attracts her to "effeminate tranny boys," Samantha muses on a kind of gender-physics that combines partners who both are "kind of femme" and yields an "energy" and "dynamics" which are "always kind of really fun." "Like, maybe it's just maybe about, like, being, like, both of us being kind of femme that's fun, you know? Like, playing off that energy." Yet Samantha refuses any sort of "fixed idea" of lesbian identity in the pursuit of "living out our gender and our sexuality" in "a different way."

> It's not some, like, clueless thing that I wouldn't be open to having, or in general, that one of us should be, like, have some fixed idea of, like, "I'm a lesbian, and I will always be a lesbian." And, like, that to me just is so, like, rigid, and like, and has all these rigid implications for, like, gender and sexuality and, like, the body, and all these ways that just aren't, to me, don't make sense, so, like, you know, and, so that to me, like, delimits, like, anything, like, any possibility, so. Yeah. And also that it's, like, political, too, 'cause that's like, a really big part of being queer for me, is, like, the politics behind it, and, like, try to practice, like, a different way of, like, living out our gender and our sexuality, and our sexual practices, you know?

We turn here to ethnographic linguistics (Hymes 1996) to understand Samantha's perspectives on sexual and gender identity in relation to her use of less coherent talk. Hill (1996; 2000) has noted that individuals narrating transgender and gender-queer identities may evidence a "double voice," representing locations both inside and outside the existing binary gender system. This "double voice" becomes apparent in Samantha's narrative through her use of the word "like" alternately as a "filler" (a word that functions to allow the speaker to gather her thoughts) and a "focuser" (an indicator that the speaker is about to say something that deserves special attention) (Petersen 2004). Substituting the word "like" for the word "said" is another common move in vernacular-a use consistently employed by Reid (below) in a way that strongly resembles African American narration of conversations (Labov 1970; 1982).

Girls have been shown to be more likely than boys to use words such as "like" to circumlocute their ideas, with explanations ranging from girls being "more comfortable speaking on the spur of the moment" (Temple University 2002) to the concept that girls try to minimize conflict by speaking less directly (Cameron & Kulick 2003). As Samantha speaks, her use of "like" generates a quicker cadence, a signifier for many that the speaker is intelligent (Tannen 2003). The use of fillers can be strategic, affording time to develop her thoughts on complex ideas in the context of the unequal relationship of informant to interviewer (Fowler et al. 1979). When used as a "hedge" word, "like" may cue that forthcoming speech is not as specific or clear as the speaker would prefer (Stern 1997; Temple University 2002). When Samantha's fillers, hedge words, and focuser words are edited out, the text becomes much more comprehensible.

> It's not some . . . clueless thing that I wouldn't be open to having, or in general that one of us should be like, [or] have some fixed idea of . . . [as in] 'I'm a lesbian and I will always be a lesbian. . . .' That to me is so rigid, and has all these rigid implications for . . . gender and sexuality and . . . the body, and all these ways that just . . . don't make sense . . . and so that, to me . . . delimits . . . anything, . . . any possibility. . . . Also that it's . . . political too, 'cause that's a really big part of being queer for me, is . . . the politics behind it . . . try[ing] to practice . . . a different way of . . . living out our gender and our sexuality and our sexual practices.

However, without the "voice" of fillers, hedge words, and focusers, Samantha's location in complexity and her experience of uncertainty are lost. In the edited "translation" above, it becomes clear that embedded within her easily dismissible pop culture/valley-girl "lingo" is her own complex thinking through the language of critical/postmodernist theory. By listening to her "likes," we can appreciate her dual location within stereotypically patterned speech (Hymes 1996) and within "a different way of . . . living out . . . gender."

Whereas Samantha considers a fixed lesbian identity as limited and limiting, "fag culture or style," or possibly some "dyked-out version of it," represents "what everybody is, like, wanting" or "wanting to identify with." Samantha tries to explain what it means for a woman to be "fagged out" by describing someone she currently dates, but finds herself at a loss for terminology.

> This girl I've been dating, she, she's like, totally, like, kind of, like, fagged out. You know, she doesn't identify as a boy or anything. But it just is like, somehow a little bit more like what, I don't know. A little bit more, like . . . God! No word to identify it, but if, like, when I see it, I know it (laughs). And, like, yeah. It seems like everybody these days is, like, trying to, like, emulate, like, fag culture or style in all these different ways. Like, in this, like, dyked, dyked-out version of it, it's really just, like, different, but it seems like what everybody is, like, wanting to do, and, like, wanting to identify with or something. I don't know.

Rather than being something available to description and critique, "fag culture or style" is something recognized implicitly and visually. For transgender "boys," fluid gender with no "demarcation" meets brand-name slacks and shoes in a "femme-y" masculine look with "continuity."

> I feel like, I feel like every, I feel like all of the boys I go out with always have a pair of charcoal gray pants, and, like, really stuck through that, like, black Chuck Taylors. Not like that's what it's all about, but there's, that continuity has always been there. The charcoal pants and the Chuck Taylors (laughs). Just totally random. I don't know, like, effeminate boys who, like, do their masculinity in a way that's really, like, femme-y. You know, they're not, like, butch and tough, or whatever, you know? I think that's really sexy. Trans boys. I mean, I'm not, like, opposed to dating anybody. Like, there's no, like, like, demarcation. But I usually end up dating trans boys.

Whereas Samantha considers tranny boys to look "preppy," "fagged-out girls" look "arty."

> Just kind of like, arty and like, she's totally, it's like, that would be, like, simultaneously, like, a boy, like, not like me, like, really girly, but like a boy, but also, like, kind of, like, effeminate, kind of?

And it's always been, like, you know, she's not like a top, like, she wants, you know? But still wants to be on the, like, I do the boy things in the relationship, but, like, I'm not, like, you know. They're all, like, always really nice, too. Whatever. You know, it's not about being, like, really a top and strong in one particular way. It's about like, something else.

Compared to "femme-y boys" who are defined through their look, fagged-out girls are defined largely through their social and sexual roles. "She's not like a top"–rather, "she wants," while letting her partner "do the boy things in the relationship." Most important, fagged-out girls are "all, like, always really nice, too."

Samantha's intricate social mapping of types of girls and types of tranny boys and their distinguishing characteristics forms a classification system that begins to drift from "other" aspects of lived experience. When asked whether there were any other topics the study should address, Samantha responded,

Well, I don't know. Like, you're studying the qualitative part of stuff, you know. What about things that, like, physical health, and, like, eating? Yeah, I guess that stuff. I don't know.

While interested in adding topics relevant to her physical health, Samantha felt that the topic of her own development was "off-limits."

Ethnographer: And your personal evolution? Is there something that we should be, the study should look after, or . . . ?

Samantha: Too weird to think about! (laughs)

Thus, by listening to queer-identified youths' discourses about queerness, and asking youth how to extend the range of their own narratives, it becomes possible to also identify areas of inquiry that have not been actively linked to queer identity (i.e., "normal" topics like physical health or disordered eating) or that have been considered "too weird" (such as self-development) to include in an LGBTQ research agenda.

REID: "I CAN NAME ALL THE DIVAS, OLD AND NEW"

Reid, a 20-year-old homeless youth who grew up in a working class family in the South, recently moved to New York City to locate within an accepting LGBT community. He considers himself "transgender queer" and a "trans-man," but has not engaged in any forms of medical transitioning (i.e., hormones or surgeries). His housing program found him an apartment in Brooklyn, where Reid buys his marijuana from African American dealers on his block, socializes and "smokes up" with youth of color "and a few white friends," and takes pride in being "evolved white trailer trash, because I'm not prejudiced, even though I was born (white)." He has recently begun to moderate his tobacco smoking, citing chronic asthma and "some kind of lung thing" that does not seem to be relieved by the inhalers a friend pilfers from a local asthma clinic. Subsequent follow-up interviews have served as platforms for Reid to discuss a range of issues relating to transgender men, to reflect on his past and present relationships to "butch lesbians," and to grasp for language to describe himself and his "soul searching" about gender transitioning.

When asked who is the most important adult in his life, Reid answers in terms of entire staffs of gay youth service programs: "The GO! Staff. And the Coming Home Staff." The GO! youth services program is a source of support services and activities for an increasing number of LGBT youth, some of whom live in apartments obtained through one of the two local programs providing temporary housing for LGBT homeless youth. Streetwise from years of transient living in several East Coast cities, Reid lives by the GO! Program's "rules," summed up in the phrase "Crab's Ass!"

> That's Confidentiality. Respect. Attentiveness. Be open. Sensitivity. Then, 'ass' stands for 'assumptions.' CRAB'S ASS!

At the GO! Program, Reid explains, youth are also exposed to "things like gender issues, like identi–, okay, iden–, identifying yourself in oppression and all this good shit."

Similar to Samantha, Reid is dedicated to uprooting assumptions about gender and sexuality. His first real "encounter" with the notion of transgender men caused him to confront some of his own assumptions about lesbians and being lesbian. While "smoking up" marijuana with other LGBT friends at the apartment provided through his housing program, Reid first learned about transgender men from Angie, a homeless youth who Reid considers to be "a gender chameleon."

> We were, okay, smoking up, cause I started smoking up with them every now and then, and we have the most intelligent conversa-

tions. So I was, like, "What is, what's your definition of transgender?" 'Cause I'm always hearing about M to F. And she's like, "Well, it's where you have M-to-F, F-to-M." I'm, like, "Wait a second. I never heard the term F-to-M before." I'm like, "What is an F-to-M?" And she's, like, "It's a female to male." I'm like, "I didn't know that could be done." You know? I'm like, "All I've heard about is M-to-F." I'm, like, "Tell me more." And, 'cause she herself identifies as transgendered queer. You know, she's a, she is an F-to-M, but she's not gonna change her body. You know, she's just gonna stay queer, 'cause she thinks, acts, like a male, even though she dresses like a female. She's a gender chameleon. You know, she can do either/or.

Reid respects Angie for deciding to "stay queer," and has made only minor modifications to his gender presentation since adopting a male name. By opting to remain in an extended "transitional" state, Reid resists adopting a "unified" male self-presentation, rejecting the medicalized norms of gender transitioning (Fuller, Mauk & Welle 2005) and generating a "queer" dimension and meaning to his embodiment as a transgender man. As an "either/or," Reid remains on the margins of the emerging community of transgender men and the "gay community" in New York City.

Reid's understandings of transgender men were conveyed through a comparison with butch lesbians, emphasizing the difference between "*dressing* like guys" versus "*feeling* like a man" and "*thinking* of oneself as a man." He recalls his moment of epiphany:

"Wait a second." I'm like, "So all these years, I felt that I was a butch lesbian, that all butch lesbians thought of themselves as men. You know? And wanted to be men. I thought that was normal. It's not?" She's like, "No. Butch lesbians are female, they just dress like guys because that's how they're comfortable. They don't feel like a man, they don't think of themselves as a man." She's like, "And you do?" I'm like, "Yeah. That's the way I've always felt." And she's like, "Wow." She's like, and I'm just like, afterwards, I'm sitting there, I'm like, "Angie, I'm transgendered."

Of particular significance is the fact that Reid had his own prior idea of what a "normal" lesbian was; the idea that "normal" lesbians "don't feel like a man" and "don't think of themselves as a man" was central to his identity shift to transgender male. This need to integrate a sense of "being queer" and "being normal" (in a double sense of being true to some notion of a core self and fitting in according to community standards) infuses many of the narratives of queer-identified youth and transgender men in the study cohort.

Reid's narratives about being "transgendered" are embedded in family memories and family relationships. When Reid told his brother that he was transgendered, his brother recalled a childhood memory of watching a show together about hermaphrodites. Reid retold the story of how, while their parents were out bowling,

> We watched this show on intersex, hermaphrodites, and I was watching about this girl who had her penis cut off and was just left with her female organ. And she was going on, claims about how she was a male, how she was a man, it was fucked up that her parents did that to her. And they described how ways you could tell someone could have been intersex. And I just watched it, and it's like, you're tomboy, they even think of themselves as a male, and whatnot. So I called up my grandmother after the show, my brother's sitting there watching. Crying, I'm like, "Why did you guys cut off my dick? You know, I'm a man!" You know, you know, crying. I'm like, "Was I a hermaphrodite? Did I have both sexes when I was younger?" She's like, "No, you were born female." So that must be, like, "Okay, well, maybe I'm just cracked."

Whereas Samantha's family "overlooks" her "gayness," Reid's family serves as a keeper of family memories, a co-chronicler of Reid's transgender history.

Unable to confirm a biological narrative about being born intersex, Reid rooted his transgender "origin story" in memories of learning "manly things" from his grandfather.

> My grandpa is the one who taught me how to be a man from as long as I was little, you know, he would teach me manly things, so he's like, "I'm glad you came around. I knew it before you were, you know, you know." . . . 'Cause when I was seven . . . he'd teach me how to shave my face. And he bought me Old Spice when I was a kid. I used to get into his, and so he bought me my own.

In addition, Reid's coming out as transgender to family members resulted in a subsequent shift in perceived familial roles. After Reid came out to his younger brother,

> He's like, "I can't call you mom anymore." He's like, "Now you're my big brother, you know, I'm gonna call you dad from now on."

While anchoring his transgender identity in social relationships (specifically, familial ones), Reid also referenced his biological heritage through the men in his family.

> I know once I go on testosterone, I'm gonna turn all muscular 'cause of a, 'cause the guys in my family are just naturally tough. . . .

Thus, once again we see an example of interfacing "queer" or "transgender" identities and heteronormative or "naturalized" identities as they reference biological or familial heritage.

Perhaps the most complex matrix of processes facing LGBTQ youth is their simultaneous embracing of difference and their search for belonging within the LGBTQ community. Once individual youth have crafted their own individual identifications within a locally recognized set of options, they may participate in a rigid enforcement of group conformity. Challenges to group conformity can generate not only competing social groups, but also individual confusion for those who have only recently "solidified" their self-presentations and defined the directions of their attractions.

In our ongoing research, it is apparent that LGBTQ youth evidence diverse views about "who transgender men are" and that significant conflicts arise in the attempts to define and, as Samantha would say, "delimit" the relationship(s) of transgender men to lesbians and to the wider "gay community." However, whereas Samantha was concerned about keeping open the range of "possibilities," contrary to popular stereotypes many street-involved LGBTQ youth may be fiercely focused on defining a given set of "normal" LGBTQ identities and "normal" LGBTQ sexual partnerings. For example, Reid described his more recent experience of "being harassed on The Pier" for breaking what he calls

> this whole unwritten law of butch/femme dynamic. If you're butch or dressed like a man, you're supposed to date a femme, not another butch, and me and my boyfriend are both tranny boys, and we date each other. So they're like, "Are you two AGs[7] dating?" We're like, "No, we're tranny boys." "Aren't you supposed to be dating women, aren't you supposed to be straight?" 'Cause there's

this unwritten law that if you're a transman, you should be straight. You can't date other tranny men.

Implicit in many (but not all) of his peers' responses to his gender and sexuality transformations is an unstated acceptance of Reid's decision to identify as transgender, but an explicit discomfort with the range of his sexual attractions as a transgender man.

When asked who writes these unwritten laws, Reid explained how heterosexist assumptions in "society" get imported into "gay culture," an inclusive term in Reid's vernacular.

It's programmed into our heads, and the way we're born. The thought that men are supposed to be with women, and everything, and we bring it into the gay culture. That's why you have the really butch lesbians dating femme lesbians, tranny boys, F-to-M, dating women, you know. There *are* queer tranny boys. I'm one of them, my boyfriend is one of them. . . . We're gay, let's just accept each other as is, you know.

Thus, in many LGBT youths' efforts to formulate and normalize their own sexual identities and gender styles, the notion of "queer" (as well as transgender and intersex identities) may not come into play or may be perceived as threatening or disrupting to recently achieved "stabilized" identities receiving affirmation in "established" LGBT communities: a dynamic similarly evidenced in adult lesbian and gay communities' rejection of transgender, queer, and intersex people.

Although his peers may perceive his sexual attractions as confused or confusing, Reid organizes the variety of his attractions through the term "butch."

Maybe it's because they were once women, and maybe it's because they're just like me, but there's something sexy about tranny boys. You know, it's about the *equal*, you know, I'm really attracted to women, I'm really attracted to tranny boys, I'm really attracted to butch lesbians. I'm really attracted to butch! It's all the same thing to me, you know. We're all humans, we're all gay. So why can't we date them?

Defining a notion of "equality" as "sexy," Reid brings a number of identity categories–tranny boys, women, butch lesbians–into his vision of equality and into his range of attractions.

Reid also sometimes "identified [himself] in oppression" by identifying with the ways in which women are subordinated or exploited. Four months after the baseline interview, Reid reported that he and his boyfriend–"one of the most confusing creatures in the world"–had broken up. He commented, "I'm never gonna understand boys. They like to play mind games with us, the women [laughs]." Reid also identified with women in empowered roles, citing his counselor–"a very hot butch lesbian"–as his greatest role model.

> I want to be just like this woman. She's a lesbian. I just wanna be like her, she's the only counselor I've ever had that helped me. . . . I love talking to Chyrell, Chyrell is the shit. She listens! She cares. She talks to me. She's not shy, just tell you how it is. If you're fucking up, she'll tell you, and she'll be blunt honest about it, she's a Scorpio.

As Reid continued to socialize and sexually partner with tranny boys, gay men, and lesbians, he synthesized images of gay men and butch lesbians to describe himself and his ways of being.

> I am just such a fag. I'm a fag. I can name all the divas, I could sing all the divas. I'm just like, "Mariah!" "Nat King Cole!" I know all these, like, old, all the old divas, new divas, like, Britney and Madonna, Christina, I'm such a fag. But then I also have the lesbian music tastes. I'm just turning more faggy, like, even with the way I dress, come on, look at me . . . a fag mixing with some dyke-ness. I don't fit into one box. A lot of ways I'm a fag, in a lot of ways I'm a big bulldyke.

In the aftermath of the breakup with his transgender boyfriend, Reid began dating a woman who doesn't exactly "fit" into the locally accepted categories.

> She's a lesbian, typical dyke. Dyke-next-door back home. The dyke next door! That's what she is, for back home. She's like the dyke that everybody wants back home but no one here in New York really wants. That type of lesbian. . . . It's just, she's wholesome. Doesn't wear makeup. Dresses up pretty though. Feminine but yet not.

Thus, unlike some LGBTQ youth who either emphasize their "queerness" or emphasize their conformity to "typical" lesbian and gay identities, Reid incorporates seemingly contradictory categories–"queer" and "typical," "transgender" and "lesbian"–into his frame of reference. However, in describing this "mix" of identities, Reid does not reflect on the fact that his sexual partnering with males may place him at risk for HIV.

In fact, Reid continues to do "deep soul-searching" about whether to engage in medical transitioning processes, including taking testosterone and having gender reassignment surgeries. Central to his decision to "wait until [he is] ready" is Reid's concern that once he begins medical transitioning, he will no longer be attractive to or "wanted" by lesbians.

> Once I do all this, and then start taking testosterone and what-not, I'm not gonna be able to date lesbians anymore. Come on, what lesbian is gonna want to be with a trans-man? You know, it's gonna be harder for me to find relationships and what-not. It's gonna be a lot, hard. And so hence I've gotta make sure I'm ready for a giant step like this. Who knows, there's a lot of trans-men that don't change their bodies.

Thus, while he is confident in his own ability to synthesize "transgender" and "lesbian" categories and partners, he worries about being rejected by lesbians once his own gender presentation and physical features are more overtly masculine. The fact that he worries about this also indicates his isolation from transgender communities in which a number of lesbian-identified women partner with transgender men.

During this decision-making time, this "transition" that may or may not follow with medical gender transitioning, Reid developed a painful cyst in his breast. He apologized to the ethnographer for being "fucked up on Vicodan," "moody," "hormonal," and "in a lot of pain."

> I have a cyst on my, o-on my breast, in my breast that's bursting at the moment. It's in the, it's like processing, it's building up, it's getting ready to burst. Um, a slight possible chance of a tumor. I gotta get a mammogram done to see whether or not I, I'll probably have to get surgery on it. All that good shit. So, yeah, nothing new, other than that.

Terrified at the prospect of surgery, uncomfortable with having attention drawn to the existence of his breasts, and wrestling with a great distrust of physicians, Reid accessed for the first time a local clinic providing health services for LGBT patients. At the next follow-up interview, the cyst had resolved, and Reid continued "processing . . . building up . . . getting ready . . . to see whether or not" he would decide to start testosterone.

DISCUSSION

Queer identifications may simultaneously serve to connect youth to each other and function as a form of self-imposed marginalization. Though each case study has presented a unique young person with divergent gender and sexuality identifications, together these demonstrate that many similar health-care needs arise for queer-identified youth. As health and social service providers aim to identify and maintain standards for LGBT health care, queer-identified youth who increasingly resist discrete identity labels may challenge practitioners to develop a highly individualized approach to care for those who "don't fit into one box." Leslie Feinberg (2001) has encouraged clinicians to resist limiting their assessments and care only through male and female, M-to-F and F-to-M categories, and to include gender-variant consumers in the emerging professionalization of transgender health. For queer-identified youth acutely aware of power hierarchies and role assignments, the doctor-patient or provider-patient relationship may need open negotiation and renegotiation. The challenges to providing a "delimited" standard of care for queer-identified youth within the various constraints of managed care or privately funded LGBT community sites are admittedly daunting.

Both Reid and Samantha describe participation in sexual roles and sexual behaviors that are vulnerable to being dismissed as esoteric. The development of an HIV prevention approach for queer-identified youth requires a careful disentangling of sexual roles and sexual practices. For example, both Samantha and Reid frequently refer to "bottoming," a term from gay men's vernacular that can variously refer to the receptive role in anal intercourse or a submissive social role in public interactions. However, when used by queer-identified youth, the term "bottom" is behaviorally highly ambiguous, potentially referring to social roles as well as a range of possible sexual activities (including anal, vaginal, or oral intercourse) with variously-bodied partners. Penetrative or receptive roles may variously involve the use of a penis, hand, finger, tongue, or sex toy. For participants in S/M scenes, the term "bottoming" may refer to sexualized roles and practices that may or may not include sexual intercourse, but that may present unique risks for HIV transmission. Thus, with an eye to HIV prevention, providers and researchers need to prompt elaborations of the specific behavioral practices in which queer-identified youth engage and solicit descriptions of partners' gendered bodies, while gaining an understanding of the meanings youth assign to various sexual practices and scripted roles.

In his identity-centric developmental model, Erik Erikson (1968) never imagined that adolescents would need to learn to independently maneuver a managed care bureaucracy. There is an urgent need for "transitional support services" for queer-identified youth as they graduate from or "age out of" the relatively protective environs of the university or supported housing programs, where basic health services may be more readily accessible. Samantha and several other college graduates in the study faced their first-ever transition into workplace-based health insurance with trepidation. Reid expressed sheer elation upon learning that his long-term counselor would continue to be available in Reid's final months of supported housing and the transition to employment and seeking his own apartment.

Queer-identified youth generate unique challenges to the framing of HIV prevention and other health education initiatives, particularly since nonconformity–including "fluid" partnering practices–may be a vital expression of queer-identified youths' politics and selves. Both Reid and Samantha could be considered "behaviorally bisexual" (although neither identify as such) and could benefit from HIV education that addresses the needs of individuals who sexually partner with transgender and birth-assigned males and females. Similarly, both may need access to gynecological care, including birth control; in this context, the specific and multiple meanings individuals assign to condom use are relevant. While risk for pregnancy and HIV infection may vary within different partner dyads, risk of sexually transmitted infections (STIs), especially Human Papilloma Virus (HPV), may be high. Environmental health may be of particular relevance for queer-identified youth who obtain housing in marginalized urban neighborhoods. Reid, who reports chronic asthma and a reliance on bronchial inhalers, resides in a neighborhood evidencing among the highest asthma rates in the nation.

Diverse discourses are employed and synthesized in the communication of queer identities (Wortes, Adams, Ziek & Welle 2005; Fuller, Mauk & Welle 2005). For queer-identified youth seeking health or support services, it is essential to be able to define themselves (and their sexual partners) through the terminology (e.g., academic vs. "Pier" discourse) of their choice, which itself may be "in development." Since youths' language practices may be highly localized and meanings highly individualized, service providers can benefit from adopting an anthropological approach: aiming to identify group-level identity terminology and to understand language uses and meanings at the individual level, in life history context. Providers serving queer-identified youth need to be able to listen non-judgmentally to academic discourses and "pier/peer" discourses about identity, and "hear" references to individual needs that may be embedded in assertions of self-reliance and critiques of professional practice ("CRAB'S ASS!"). For providers located in LGBT community sites where community or agency politics inform institutional practice, a less openly ideological approach to queer-identified youth may be particularly facilitative. Queer-identified youth share along with other adolescents the project of forming a self both with and against societal or community-level standards and expectations, and may risk the loss of essential supports in rigidly ideological mainstream or LGBT service settings. At the same time that queer-identified youth may reject heteronormative standards of self-identity and interpersonal relations (Nestle et al. 2002), providers need to engage queer-identified youth in both individualized and routine health services, and assist youth who experience health challenges that may be easily associated with heteronormative relations, including partner violence, rape, pregnancy, or abortion.

While calls for research on adolescent development among "sexual minority" youth continue (Savin-Williams 2001), the task remains to elaborate a model of adolescent development that can adequately address the unique complexities of LGBTQ youth through a dual focus on their formation of sexual and gender identities and their ego development in contexts of community-level and institutionalized intolerance of sexual- and gender-variant individuals. Queer-identified youth provide important insights into developmental complexity among LGBTQ youth, underscoring Noam's (1996: 7-10) observation that "more complexity" may be accompanied by unique vulnerabilities. As our case studies suggest, queer-identified youth participate in developmental complexity through rejections and elaborations of recognized genders and sexualities, through the dual formation of "queer," "fluid" identities and "normal," bounded selves, and through the prioritization of "identity talk" often at the cost of silencing the body and needs for self-care. It remains for health researchers, providers of LGBTQ health services, and empowered local communities to identify ways to encourage queer-identified youth to "indulge" in the "taboo of taking care." For each generation, it seems, health becomes a queer pursuit indeed.

ISSUES AND QUESTIONS FOR FURTHER DISCUSSION

1. What does the identity label "queer" represent? How can you see the definition of "queer" change according to the socio-cultural context of the user? In the article, how do Reid and Samantha use the term "queer" to describe themselves and others? What are the differences/similarities?

2. In the article, Samantha describes developing her "queer politics" through rejection of "liberal feminism" and feels "thankful" that she wasn't "influenced" by feminism (p. 9-10). How does Reid construct certain lesbians in his life as "role models," yet retain a notion of himself as "queer"? What are the implications of queer identity constructed as pure and superior versus synthetic and accomodating?

3. How does adopting an identity that is formed in rejection of other, more fixed identities, serve to isolate or connect the people who identify this way? Can you envision a context in which Samantha and Reid might occupy the same social space?

4. Are there aspects of your own everyday vernacular in which gender is used as a fixed concept? A fluid concept? What influences the decisions to use gender in these ways?

5. How do you think sexual self identity influences accessibility to health care? Do you think Reid and Samantha access the same health care service providers? Why or why not?

6. What are some questions that you think a physician could ask Reid in order to facilitate an open physician-patient relationship? What are good questions for Samantha?

AUTHOR NOTE

This research was funded through a grant awarded to the first author by the National Institute for Child Health and Human Development (Grant R01 HD 41723-01) and support from National Development & Research Institutes, Inc. Views expressed here do not necessarily represent those of NDRI or the U.S. Government. We acknowledge the contributions of fellow members of the 2Grown Project, Damaris Wortes, Sherry Adams, Glen Barnard, and Kristine Ziek, and are especially grateful to Samantha and Reid and their peers for their ongoing participation in this project.

NOTES

1. Personal Websites (e.g., *http://kpscapes.tripod.com/index.html, http://friffboy. tripod.com*) feature autobiographic writing accompanied by photojournalistic reporting of medical transition, and e-mail listservs host online discussions (e.g., *http:// groups.yahoo.com/group/ftmdisc*). Informational Websites (e.g., *www.transter.com*) show "before and after" surgical pictures and list testosterone effects (*http://tghealthcritiques.tripod.com/ftm_ha1.htm*).

2. (Meyer et al. 2001) (i.e., the controversial medical and mental health treatment protocols for gender transitioning).

3. Ethnographic research utilizes anthropological methods, including life history interviews, open-ended interviews, and participant-observation ("hanging out" or immersion in social scenes). The dual emphasis on capturing "variability" of social practices and individual "voice" (Clatts, Welle & Goldsamt 2001) illuminates the intersection of behavior and identity. See Geertz (1973), Mead (1973) and Agar (1980). For an introduction to LGBTQ ethnography, see Herdt and Stoller (1990), Weston (1991), and Newton (2000).

4. Gil Noam (1996) has elaborated a model of developmental complexity that emphasizes the traumatic nature of human development and the intrinsic role of loss in informing progressive development. Noam challenges classic frameworks which pathologize age/stage dysynchrony, emphasizing instead the protective and potentially productive nature of developmental delay, and noting how children and youth test new skills and acclimate to new perspectives during such durations. This can inform an understanding of how individual youth "time" their own gender and sexuality disclosures. Noam emphasizes the importance of biographical narrative and the production of meaning in human development. While these contribute to the complexity of individual perspectives on the self and the world, complexity itself is not necessarily protective or indicative of "better mental health." Hence the need to attend to both the sources of complexity (e.g., perspectives on gender and sexual identities) among queer-identified youth and the attendant strengths and vulnerabilities of each individual.

5. "The Piers" is an LGBTQ youth venue located on Pier 45 at 10th Street along the Hudson River, in Manhattan's Greenwich Village. After the decline of the shipping industry in the mid-twentieth century, Pier 45 and some adjacent piers became a venue for gay cruising. More recently, the dilapidated concrete structure also became a place where LGBTQ homeless youth could find shelter, form friendships, and generate income via sex work. In early 2003, while this study was conducting recruitment, the ribbon was cut on Pier 45 launching the renovated grounds as part of the Hudson River Park project. Even with increased policing, lighting, and general public usage, "The Piers" remain one of the few public "hangouts" for LGBTQ youth, particularly LGBTQ youth of color who commute from the outer boroughs, upper Manhattan, and New Jersey.

6. In the 2Grown cohort, nearly half the participants provided unprompted descriptions of subclinical disordered eating at some point in their life, with many reporting having engaged in risky eating behaviors while planning to come out or right after coming out. We are planning future research to examine the ways that disordered eating relates to notions of sexuality and gender among LGBTQ youth.

7. "A.G." is an emerging sexual identity label increasingly adopted by African American and Latina women. Similar to the butch lesbian construct, but distinct in its signifi-

cation through clothing styles emerging from hip-hop culture, "A.G." stands for "aggressive." Our interviews with service providers suggest that young women of color began to reappropriate the term as a sexuality identifier after being diagnosed as "aggressive" in mental health assessments. In the wake of the 2003 murder of Sakia Gunn (age 15) who was stabbed on her way home from The Piers, her friends have launched the Sakia Gunn Aggressive'z and Fem'z to provide support for young lesbians. See *www.newarknow.org/newsletter/news5.htm*

REFERENCES

Agar, M. (1980). *The professional stranger: An informal introduction to ethnography.* New York: Academic Press.

Battle, J., Cohen, C. J., Warren, D., Fergersen, G., & Audum, S. (2000). *Say it loud I'm black and I'm proud: Black Pride Survey 2000.* New York: The Policy Institute of the National Gay and Lesbian Task Force.

Bockting, W. O., Robinson, B. E., Rosser, B. R. (1998). Transgender HIV prevention: A qualitative needs assessment. *AIDS Care, 10* (4), 505-525.

Bucholtz, M., Liang, A. & Sutton L. (1999). *Reinventing identities: The gendered self in discourse.* London: Oxford University Press.

Cameron, D. (1992). *Feminism and linguistic theory (2nd Edition).* New York: Macmillan.

Cameron, L. (1996). *Body alchemy: Transsexual portraits.* San Francisco: Cleis Press.

Chiseri-Strater, E. (1991). *Academic literacies: The public and private discourse of university students.* Portsmouth, NH: Boynton.

Clatts, M.C., Welle, D.L. & Goldsamt, L.G. (2001). Reconceptualizing the interaction of drug and sexual risk among MSM speed users: Notes toward an ethnoepidemiology. *AIDS and Behavior 5*(2): 115-130.

Connell, R. W. (1995). *Masculinities.* St. Leonards, Australia: Allen & Unwin.

Cromwell, J. (1999). *Transmen & FTM's: Identities, bodies, genders & sexualities.* Champaign, IL: University of Illinois Press.

Davis, A. (1981). *Women, race & class.* New York: Vintage.

Erikson, E. (1968). *Identity: Youth and crisis.* New York: Norton.

Feinberg, L. (2001). Trans health crisis: For us it's life or death. *American Journal of Public Health, 91* (6), 897-900.

Foucault, M. (1977). *Discipline and punish: The birth of the prison.* New York: Pantheon.

Fowler, R., Hodge, R., Kress, G., & Trew, A. (1979). *Language and control.* London: Routledge.

Fuller, S., Mauk, D. & Welle, D. (2005). Gender Narratives as a Source of Developmental Complexity Among Transgender and Gender-Questioning Youth. Poster Symposium, Biennial Meeting of the Society for Research in Child Development (SRCD), Biennial Meeting, Atlanta, GA (April 7-10).

Futterweit, W. (1998). Endocrine therapy of transsexualism and potential complications of long-term treatment. *Archives of Sexual Behaviors, 27,* 209-226.

Geertz, C. (1973). *The interpretation of cultures.* New York: Basic Books.

Gordon, L. (1976). *Woman's body, woman's right: A Social History of birth control in America*. New York: Penguin Books.

Herdt, G. & Stoller, R. (1990). *Intimate communications: Erotics and the study of culture*. New York: Columbia University Press.

Hewitt, P. & Warren, J. (1997). *A self-made man: The diary of a man born in a woman's body*. London: Headline.

Hill, D. B. (1996). The postmodern reconstruction of self. In C. Tolman, F. Cherry, R. van Hezewijk, & I. Lubek (Eds.), *Problems of theoretical psychology* (pp. 265-273). North York, ON: Captus Press.

Hill, D. B. (2000). Categories of sex and gender: Either/or, both/and, and neither/nor. *History and Philosophy of Psychology Bulletin, 12* (2), 25-33.

Hymes, D. (1996). *Ethnography, linguistics, narrative inequality: Towards an understanding of voice*. Philadelphia: Taylor & Francis.

Israel, G. E., & Tarver, D. E. (1997). *Transgender care*. Philadelphia: Temple University.

Joffe, C. (1987). *Regulating sexuality*. Philadelphia: University of Pennsylvania.

Karasic, D. H. (2000). Progress in health care for transgendered people. *Journal of the Gay and Lesbian Medical Association, 4* (4), 157-158.

Kotula, D. (2002). *The phallus palace: Female to male transsexuals*. Los Angeles: Alyson.

Labov, W. (1970). *The study of non-standard English*. Champaign, IL: National Council of Teachers of English.

Labov, W. (1982). *Language in the inner city: Studies in the black english vernacular*. Philadelphia: University of Pennsylvania.

Leap, W. & Boellstorff, T. (2003). *Speaking in queer tongues: Globalization and gay language*. Chicago: University of Illinois.

Livia, A. & Hall, K. (1997). *Queerly phrased: Language, gender and sexuality*. London: Oxford University.

Lombardi, E. (2001). Enhancing transgender health care. *American Journal of Public Health, 91* (6), 869-872.

Mead, M. (1973). *Coming of age in Samoa*. New York: William Morrow.

Meyer III, W., Bockting, W., Cohen-Kettenis, P., Coleman, E., DiCeglie, D., Devor, H., Gooren, L., Joris Hage, J., Kirk, S., Kuiper, B., Laub, D., Lawrence, A., Menard, Y., Patton, J., Schaefer, L., Webb, A., Wheeler, C. (2001, February). The standards of care for gender identity disorders, Sixth Version. *The International Journal of Transgenderism, 5*(1).

Nestle, J., Howell, C., & Wilchins, R. (Eds.) (2002). *Gender queer: Voices from beyond the sexual binary*. Los Angeles: Alyson.

Newton, E. (2000). *Margaret Mead made me gay: Personal essays, public ideas*. Durham: Duke University Press.

Noam, G. (1996). High-risk youth: Transforming our understanding of human development. *Human Development, 39* (1), 1-17.

Oriel, K. A. (2000). Medical care of transsexual patients. *Journal of the Gay and Lesbian Medical Association, 4* (4), 185-194.

Petersen, A. (2004, February 21). Like it or not, "like" is leaking into adult language. *St. Paul Pioneer Press*. Available: *http://www.twincities.com/mld/twincities/living/8000492.htm*

Rubin, G. (1975). "The traffic in women," In R. R. Reiter (Ed.), *Toward an Anthropology of Women*. New York: Monthly Review.

Rubin, H. (2003). *Self made men: Identity and embodiment among transsexual men*. Nashville, TN: Vanderbilt University.

Savin-Williams, R. 2001. A critique of research on sexual-minority youths. *Journal of Adolescence, 24*, 5-13.

Stern, B. (1997, Spring). Advertising to the "other" culture: Women's use of language and language's use of women. *National Forum: The Phi Kappa Phi Journal, 77* (2), 36-39.

Tannen. D. (2003, January 5). Did you catch that? Why they're talking as fast as they can. *The Washington Post*. pg. B. 01.

Temple University/Office of News and Media Relations. (2002, August 20). Temple University professor cracks the code: Reveals meaning of the teen word, "like". *News Releases* [On-line]. Available: *http://www.temple.edu/news_media/jr193.html*

Welle, D.L., Clatts, M. C., & Barnard, G. (2005). "Not just infatuation": Sexuality and literacy in the age of HIV. In Brian V. Street (Ed.) *Literacy across educational contexts*. Philadelphia: Caslon Publishers, 261-282.

Welle, D.L., Clatts, M.C. & Vidal-Ortiz, S. (in press). "Understanding disordered eating among transgender persons: An ethnographic case study." *Culture, Health & Sexuality*.

Weston, K. (1991). *Families we choose: Gays, lesbians and kinship*. New York: Columbia University Press.

Wortes, D. I., Adams, S., Ziek, K. & Welle, D. L. (2005). "People label everything": The essential role of language practices in LGBTQ youth development. Poster Symposium, Biennial Meeting of the Society for Research in Child Development (SRCD), Biennial Meeting, Atlanta, GA (April 7-10).

"Gee, I Didn't Get That Vibe from You": Articulating My Own Version of a Femme Lesbian Existence

Robbin VanNewkirk

SUMMARY. This article aims to demystify the notion of a *gay vibe* from a femme queer woman's perspective. It contextualizes the author's experience of being read by the queer community as straight and of doing femme as a means for placing oneself on the gaydar screen while questioning the role of signifiers in creating a myth that sexuality is always concrete and permanent. This article goes further to argue that femme sexuality, because of its occasional invisible state, has the potential to move between ideological positions in order to destabilize them.

Robbin VanNewkirk is a graduate student in the Women's Studies Department at Georgia State University in Atlanta, Georgia. She is currently working on a thesis about the use of the progress narrative in third wave feminism. Always interested in the in-between or what is not said, she spends much of her academic time finding ways to describe exclusion. She also expresses this continuous desire through radical cheerleading or on stage with The Dixie Pistols, an all queer girl burlesque troop.

Address correspondence to: Robbin VanNewkirk, Women's Studies Institute, Georgia State University, P. O. Box 3969, Atlanta, GA 30302-3969 (E-mail: nvan1@student.gsu.edu).

[Haworth co-indexing entry note]: "'Gee, I Didn't Get That Vibe from You': Articulating My Own Version of a Femme Lesbian Existence." VanNewkirk, Robbin. Co-published simultaneously in *Journal of Lesbian Studies* (Harrington Park Press, an imprint of The Haworth Press, Inc.) Vol. 10, No. 1/2, 2006, pp. 73-85; and: *Challenging Lesbian Norms: Intersex, Transgender, Intersectional, and Queer Perspectives* (ed: Angela Pattatucci Aragón) Harrington Park Press, an imprint of The Haworth Press, Inc., 2006, pp. 73-85. Single or multiple copies of this article are available for a fee from The Haworth Document Delivery Service [1-800-HAWORTH, 9:00 a.m. - 5:00 p.m. (EST). E-mail address: docdelivery@haworthpress.com].

73

This form of differential consciousness has subversive potential because it renders gaydar technology unsubstantial and broken. *[Article copies available for a fee from The Haworth Document Delivery Service: 1-800-HAWORTH. E-mail address: <docdelivery@haworthpress.com> Website: <http://www.HaworthPress.com> © 2006 by The Haworth Press, Inc. All rights reserved.]*

KEYWORDS. Femme lesbians, butch lesbians, queer identity, femininity, gay vibe, gaydar, differential consciousness, coming out, Chela Sandoval, Judith Butler, Roland Barthes, myth, realness

Recently, I attended a nontraditional Thanksgiving dinner at a yoga house in my neighborhood where a lesbian woman asked me what women's studies means. She jokingly said, "I study women too, but probably not for the same reasons you do." Suddenly, I found myself knocked unsteady by this stranger's transparent assumptions about my personal sexual identity and particular location within my field of study, and there is no quick, clever response I've found for being shoved back into the proverbial closet. I corrected her arrogation along with a nervous laugh, to which she replied, "Gee, I really didn't get that vibe from you." This reassertion of my invisible social location immediately placed me outside of her understood categories for sexual identity, categories which are, as I discovered, discernable by some sort of mysterious personal radar technology. I went home and stared at my reflection in the mirror, assessing my pink sweater, black skirt, and fishnet stockings to see if the vibe might be detected somewhere beneath the layers of this femme ensemble, assuming the vibe might just be a little obstructed.

By its very definition, a vibe is not anything tangible or concrete; it's informal vernacular, slang for an *intuitive feeling*. Technically, I wouldn't be able to *see* my vibe, and yet, a gay vibe seems to have so much to do with those observable signifiers for the *authentic* lesbian. I often wonder if I could make myself appear more gay, as I seem to remain in a perpetual state of dislocation and in-between-ness. My transformation between worlds causes me to feel like I belong to a hyphenation of the sexual binary or a liminal dimension without a name that interrupts the fixed stages; I'm situated inside a gap between both compulsory straight normalcy and creative queerdom. I understand this compulsion we all have to cognitively map people's identities, but I also understand that we are limited by the language available to us, and this hurts people in cases where words seem essential to belonging. Ultimately, this vibe I apparently wasn't giving off at the dinner party might tell us less about the essence of an object and more about the history of images and messages repeated until they are believed. And yet, while it might be a noble activity to eschew fixed categories that we don't find useful or indicative of anything we represent, in the mainstream state of lesbian and gay pride that largely relies on the grand narrative of coming out and being out, what does it mean for those of us who remain under the gaydar?

I am articulating my own version of femme lesbian existence. I am articulating and rearticulating an identity with room for flexibility and forgiveness while I live outside the perimeters of people's gaydar that has the capacity to conspicuously register desire. My development as a femme is tied to a history of lesbian sexuality both compelling and provocative for its intense erotic flavor and for the oppression experienced by the femme-butch community. Who doesn't want to cry when they read Joan Nestle's edited volume of stories and essays in *The Persistent Desire: A Femme-Butch Reader*? It is useful to examine historical records of the femme-butch community, a history that is marginalized by heterosexist ideological constructs of desire and romance in order to get to the power-laden reasons for this oppression. It is this ideological construct of desire that shadowed my own sexuality, allowing me to study and prefer women while remaining ambiguously heterosexual for many years. Yet, I refer to myself as a femme with some amount of hesitation and discomfort due in part to my uncertainty of what this identification means, but also due to my moderately poststructuralist tendencies that drive me to break down anything that resembles a constricting label, something that becomes immobile and even marketable. Meanwhile, I believe that personal narration is the nectar of life–it becomes difficult to tell a story or describe an experience without reassembling these very same categories of difference that might be problematic within some theoretical terrains. The point is not to discredit experiences, but to expand the definitions and the technology that perceives difference in order to include those of us who normally remain illegible.

I refer to myself as a femme in large part because when I began to date women I either realized or was told I was a femme through discourse occurring in the lesbian community. I can now hear a voice reflecting Simone de Beauvoir, repeating emphatically, "One is not born a femme, but rather becomes one." It leaves me wondering, why am I a femme when I play in the queer community and a tomboy when I play in the straight community? I am interested in how my actions affect either label, but I am even more interested in how the labels begin to influence my actions as I feel the tug and pull of expectations and resist the power in signifying the sign. Most of all, I refer to myself as a femme because my history of sexuality and experience as a woman and a feminist in both the straight and queer community parallels other femme's experiences of dissatisfaction, frustration, invisibility and even anger. Fortunately, I have never had my feminism challenged the way that some femmes in the 1970s may have had, but I have had my sexuality challenged because I don't seem to have visible signs of queerness and because I don't think I was *born gay*, and I didn't *come out* through some momentous definitive move. In many ways, the challenge of my version of sexuality and gender requires (re)identification and *coming out* every day. As the situation regarding the dinner party illustrates, it's not just the clueless straight man who assumes I must be straight and therefore sexually available according to my femininity, but also members of my own community and their assumptions about my femininity that requires a constant re-telling or coming out.

Doing femme, for me, is an appropriation of a sign that is legitimated and understood in a queer context, something that helps to strategically place me on the gaydar screen, however, I want to differentiate *doing* from *being* femme. Differentiating the two meanings from one another recognizes the way speech is used to create a form and a myth out of labeling, and treats the category as driven by vision rather than destiny. *One is not born a femme, but rather becomes one.* Often speech is used carelessly, and yet, the meaning and assumptions that follow carry weight in the signification. Femme, as a label, carries even greater weight when it gets pulled into the discourse of *queering femininity*, as if femininity is manifested in a fixed and natural state that must then be modified or ruptured by the abnormal and therefore defiant queer impact. In the most radical approach, both femininity and femme would be something one does rather than something one is while resisting the urge to measure the subversive potential of either through polarity, claiming that they are only actively engaged while they are in opposition to masculinity or heterosexuality. It is not that I am opposed to those who feel that femme is a way of being, and that it is the true apex of their identity. It is more that femme-ness comes naturally to me when I engage it, but it is not something that I naturally am.

I resist the label of femme sometimes for the same reasons a quilter might resist the label of artist. This is particularly true when people start talking about *high femme*; versus what? Thankfully, you don't hear people talk too much about *low femmes*, but it still leaves me wondering if I can truly manage this identity. I can't help but feel like femmes are supposed to be confident and legendary creatures, not awkward and skeptical. As the invaluable Shar Rednour states in *The Femme's Guide to the Universe*, "When a high femme glides through space, everything halts as the experience is absorbed" (2000, 4). Frankly, I don't glide through space so much as teeter unless I am concentrating hard on something, and then I inevitably end up tripping over some obstacle in my path. Indeed, performance of gender has become a definitive topic for queer theorists. In my own little private joke, I often think that if sexual identity and gender are such a performance, I really need to get a better agent. This is not meant to be a self-deprecating depiction of myself; on the contrary, it is a statement for the nonperformative aspects of being, the accidents that inevitably affect any intention of construction, behavior without an audience or at least an audience to validate the performance. Performance gives form to the formless and words to the speechless. Subversion interrupts a flow, a continuance that contradicts the dissonance of ends and beginnings, pauses in the motion that reflects reality. I say all of this as a strong believer that everything is a performance. Like other academics, I have been greatly influenced by Judith Butler's understanding of performativity as a sustained set of manufactured acts that are not necessarily internal essences of the body, but are rituals that are repeated and normalized (Butler, 1999). Proponents of the performativity of gender often highlight the subversive implications of queer identity, but where does that leave those of us who don't *glide through space*, but are just walking to get somewhere whether there is an audience or not? Can I still be subversive if my actions are not always a manipulative and tactical strategy for resistance? What if the subversive potential of femme identity becomes an expectation that I cannot always fulfill?

Biddy Martin points out that with queer theory comes the assumption or argument that to be truly subversive you have to be *performing gender*, that this is often interpreted as campy and intentional, and often the act of cross-dressing (whether it is the masculine butch or the high femme dressed like a drag queen) is the only potential for subversion. She states, "Such crossings have the potential to destabilize and collapse problematic boundaries. But we have to be wary of the tendency to make sexuality the means of crossing, and to make gender and race into grounds so indicatively fixed that masculine positions become the emblem of mobility" (1996, 79). Visible butch codes of identity are easily interpreted as gender performance if femininity or femme is taken for granted as the natural and standard behavior for the female body. So normalized, in fact, is the idea of femininity as a natural reflection of a deeper psychic core of the female body that it becomes difficult to try to even make clear the subversive potential of femme identity. However, femme identity does not necessarily come about unhitched to a radical consciousness that fights and undermines the social system of rewards and punishments that powerfully regulates strict expectations of how gender is supposed to be exhibited.

Subversion in its disruption of this ritual of gender performance seems to require a specific kind of understanding of the mainstream hegemonic ideology the practitioner is up against. This is often conveyed through *intellectual* terms and structures while further obscuring speech and sites existing outside the intellectual activist and academic purview. It also tends to ahistorically overlook the way that race and class have intersected with sexuality at sites where assimilation might be the means of survival or where performing gender has occurred and currently exists through different kinds of social domains.

On the other hand, subversion is trickier than simply undermining dominant categories because we often fall into the trap of reinventing those categories we wish to deconstruct. By *coming out* I am subversively undermining people's assumptions of what it means to be queer, but I am also reinventing a closet that I have *come out of.* I am reinstating the very thing that I wish to undermine and redeploy, that being heterosexuality, which becomes re-centered every time I clearly state what it is that I am not without the words to describe accurately what it is that I am. The revelation seems to take place only because heterosexuality exposes its own deviancy. In addition, while the closet is a clear image and institution, it is never fully clear what it is that one comes out into (Butler, 1996). Here, the fresh, the newer and more improved version of the self must obtain within this strict vision of the public/private spheres, meaning and structure according to precarious substandards.

Butler asks, "If a sexuality is to be disclosed, what will be taken as the true determinant of its meaning; the phantasy structure, the act, the orifice, the gender, the anatomy" (1996, 374)? If an intuitive feeling a stranger at a holiday dinner party gets based on the visible codes I embody is enough for her to closet me, what is it that becomes the true determination of my desire? Can words alone be enough for constancy and articulation of one's sense of self? Aspects of ourselves that are not necessarily definable due to the structure of language that constantly imposes difference are at the mercy of a system of words that might only paraphrase what we actually feel. Thus, anything unsteady becomes marginalized at best and uncharacteristic or unreal at worst. Bisexuality is sometimes reduced to *bi-curiosity*, and lurking behind the tone of this description is the sense that bisexuality is not a trustworthy or *true* form of sexual expression. The idea that eroticism can blur and transcend binary structures is not easily comprised by language. The homosexual/ heterosexual binary forces people into making a claim based on performance that has little to do with the complicated ways that people embody or experience desire, and interstitial spaces become closed and unrecorded. I am more interested in what cannot or will not be said. "I would want to assert unequivocally, however, that what in the mainstream culture is deployed as the definition of specific identities has no meaning until the 'other tongues' have spoken," Carol Boyce Davies argues (1996, 340). These "other tongues" imply the limitations of language to fully convey any essence or stability of any category. In some ways, the statement "Gee, I didn't get that vibe from you" realizes a certain degree of formlessness, as a vibe is not anything necessarily *there*, but with the statement is the implication of fraudulency if form, as unstated, has no real proof. If the vibe is enforcing presumptions based on only a few voices, then it hasn't relied on "other tongues" and is an obstructed lens through which to view the world. Maybe the vibe is manifested through other expressions that have not yet been identified. Or, as Davies illustrates, maybe it has a great deal to do with location. She states, "My titling of these assertions of speech, between the speaking and not-speaking space, as 'other tongues' locates speech within the context of gender, identity, sexuality and the politics of location. For it is location which allows one to speak or not speak, to be affirmed in one's speech or rejected, to be heard or censored" (1996, 340).

If heterosexuality is a construction and homosexuality disrupts this material, what do we do with notions such as the *gay vibe* that seems to suggest a natural form which gayness must take? What does it mean upon entering a community that celebrates the disruption of heteronormativity when this same community sometimes fears or doesn't see my version of queer existence? I put the question this way because of the insecurity I feel when I am not deemed real enough to be gay, and because of the gratification I feel when I can pass for gay, again with the underlying assumption that in passing I am not intrinsically real. The fact that what is experienced as *real* is conditional according to personal experience and cultural context has no bearing on prevailing myths that dictate the metanarratives within the gay and lesbian community. It might be useful, then, to try to understand the gay vibe, not necessarily in its allusive form, but as a materially constructed force and apparatus when in conjunction with personal radar technology.

In Roland Barthes's work *Mythologies,* he explores the answer to the question, "What is myth, today?" I use his work as a method for understanding how people come to believe that a gay vibe exists, and to see how this speech has taken shape within the gay and lesbian community. In this book Barthes states, "It can be seen that to purport to discriminate among mythical objects according to their substance would be entirely illusory; since myth is a type of speech, everything can be a myth provided it is conveyed in discourse. Myth is not defined by the object of its message, but by the way in which it utters this message: there are formal limits to myth, there are no 'substantial' ones" (1972, 109).

Myth is composed through a system of signifiers and the signified that decompose into the confirmed sign. As Barthes explains, when myth becomes a form, it is left behind when it no longer retains its own history. So myth is at once present and full and also distant and empty. It is similar to the way a camera, activated by a photographer, is simultaneously present but relies on its own absence to produce an objective translation of the message. While the photographer picks out a subject and manipulates the lens s/he removes all other information from the scope. The image, which is then produced, is both full of meaning, but also deprived of context when the camera is no longer a part of the composition. While it is known that a camera has produced this image, the technique, the style, the mood, the very event being recorded at the time in which the image was snapped are no longer available. Likewise, if we were to study the camera itself, the design, the materials it is made of, and how it works, we would be able to gather a basic definition of what the camera is. A complete definition of the camera would not, however, be drawn from looking at the picture it takes alone. Given this understanding of the myth, gay identity interpreted by a vibe is merely a sign signified by the signifier. Particular images and speech that have been repeated enough times to be rendered queer are registered by the gaydar and made real. Race, ethnicity, class and disability affect the power and degree to which these symbols are able to successfully convey the message. Had I been talking to the woman at the party from a wheelchair, she probably would have had her assumptions even more disrupted, as my femme sexuality would have been not only illegible, but also subsumed under another myth that people with disabilities are not sexual. The success of the signifiers also depends a great deal on the absence of the technology in order to render the sign genuine, natural, and believable in its stagnation, as if the sign exists un-codependent and without context.

The gay vibe also mystifies any possibilities of ambiguity; built into it is the inference of a binary opposition, as I become what I am not. The statement, "I didn't get that vibe from you" has meaning only as long as the opposition of gay and straight functions to point out and impose difference between the two. Sexuality as formless and without shape is something one engages, but as a form and determinant is something one is. It wouldn't make sense linguistically to make a statement such as "She is sexuality," but as a sign that is signified by a signifier it decidedly becomes a label to be represented. Queer sexuality is then determined by signifiers, which are coded more real oftentimes when they are associated with butch or female masculinity, but are oftentimes questionable when they are femme or feminine. It could be argued that gay identity, viewed as a sign or myth described as a form could indeed be cognitively registered given this understanding of how the form propels the message, but the sign is limited in its capacity to distinguish innate or permanent differences. In other words, I could wear a T-shirt with a big gay rainbow on the front that would mark me with lesbian identity and propel a gay vibe, but that wouldn't make me intrinsically more queer than anyone else in the room. It also wouldn't illustrate any true permanency for my identity, which is what I believe is most discomforting to people who want to believe that the visual codes suggest stability. It cannot accurately account for radical border-crossing or landscapes where multiple identities may converge. With this in mind, it could be suggested that what is deemed *real* by the technology of the gaydar ought to be reconfigured and expanded due to a malfunction in the apparatus. If I do not register on the gaydar then the technology is broken by its very rigidness and inability to register complexity.

Earlier I asked what the implications are for those of us who remain under the gaydar when gay pride relies largely on visibility. The categories that are signified and thus deemed legible and real mystify movement that occurs within the interstitial gaps, places where those of us sometimes remain when we are crossing borders. Butler states in *Gender Trouble*, "The more insidious and effective strategy it seems is a thoroughgoing appropriation and redeployment of the categories of identity themselves, not merely to contest 'sex,' but to articulate the convergence of multiple sexual discourses at the site of 'identity' in order to render that category, in whatever form, permanently problematic" (1999,163). In *Methodology of the Oppressed* Chela Sandoval describes the method of differential consciousness that U.S. feminists of color have strategically enacted as an activity that allows movement "between and among ideological positionings considered as variables, in order to disclose the distinctions among them" (Sandoval, 58). She states, "Differential consciousness represents a strategy of oppositional ideology that functions on an altogether different register. Its powers can be thought of as mobile–not nomadic, but rather cinematographic: a kinetic motion that maneuvers, poetically transfigures, and orchestrates while demanding alienation, perversion, and reformation in both spectators and practitioners" (Sandoval, 44). It is something that is learned by oppressed people as a survival skill. Ultimately, everything I do, including femme identity, is influenced by a feminist praxis that seeks to unveil and name power structures in order to resist them. At the same time, however, resistance occurs at various sites and in differing degrees according to the need. It is useful to recognize different tactics of resistance enacted in communities in struggle to make the mechanisms for social change even more effective, while recognizing that some of the most effective sites of resistance will always remain invisible. In fact, they are powerful in large part because they skillfully remain under the radar. Sandoval creates a method for resistance in which practitioners "intervene in reality for the sake of social justice." This technology, in its capacity to utilize multiple mechanisms, requires that the practitioner become acquainted with all the possibilities of resistance in order to effectively undermine any single dominant manifestation of power. It takes into account that agency occurs on multiple levels and through various domains and structures without shadowing the systems of oppression that make the resistance necessary.

I make the connection to her method for two explicit reasons. First, my version of femme sexual identity that requires constant (re)identification based on its instability within the mainstream lesbian discourse has the potential to strategically move between ideological positions, subversively rendering them problematic. In this case, identity, which is threatened every time it is misunderstood and pressured into becoming either permanently isolated or absorbed into something else, functions both inside and outside of the official version. Thus, it does become a perversion or reinvention of the official version while also making that version visible. Any ideology or myth that makes conforming to something *real* the only means for belonging creates a limited site, as the *real* always shifts according to the contextual information available. The gay and lesbian community risks mirroring heterosexist hegemonic power structures by limiting its membership to identity based on strict notions of the sexual and gender binary that do not allow any shifting of or transformation between sites. Differential consciousness, rather than conforming to strict standards of both belonging and resisting, is a more radical form of existence; it creates a unique space for cognitively mapping the world because it recognizes that distinctions, while useful in cases for self-preservation or narration, can also order structures and become tools of domination. Remaining under the gaydar offers potential to break free from the narrative that constricts movement and flow between ideological constructs. Differential consciousness requires, according to Sandoval, the practitioner's ability to read the current situation of power and self-consciously choose which ideological position to utilize. Thus, it is a conscious subversion and is a conscious performance of sorts. The practitioner of differential consciousness reads the technologies of power as such, and sees ideological constructions even when they are queer and subversive in their own way, as potential possibilities for resistance rather than the only possibility.

The second reason for this connection to Sandoval's method, which is inspired by resistance within "uncharted psychic terrains," is that having my sexuality challenged, and resisting the technology of power that reads me as straight, has offered me growth in my understanding of other spaces of marginality, spaces where sex, gender, race, class, and disability play a part in how people form their own versions of what is real. I see this as an advantage even though it came about through a feeling of not belonging. If being told that I don't have the gay vibe causes me to feel like a freak, I understand others whose positions of marginality may also cause them to feel like divergent outsiders. If my sexuality, by its constant rearticulation transcends stable configurations places me in an altogether different space, then I can't help but identify with others also caught in a trans world and liminal dimension. It is the tension inside of this gap that ultimately inspires all my academic and activist work. Sandoval examines the scapes that are trans-formable and views their problematic cross-positioning as a means for resistance. The consciousness in opposition that U.S. third world feminists developed intervened in the dominant power structure because of this shared experience of dislocation. She asserts, "This theoretical and methodological compass was developed, represented, and utilized by U.S. lesbians and feminists of color because, as lesbian Native American theorist Paula Gunn Allen put it in 1981, so much has been taken away that 'the place we live now is an idea'–and in this place new forms of identity, theory, practice, erotics, love, and community become imaginable" (2002, 24). In light of the fact that my identity often exists outside of the perimeters of people's ability to cognitively map me, and as I have struggled to find ways to completely express my version of sexuality, I find spaces that activate a transcultural and transgendered consciousness to be the most radically interesting. These spaces are the most effective in imbricating race, class, sexuality and gender so that they intersect to form radical subversive consciousness.

Attempting to lay out a critical inquiry of the gay vibe has returned to me a sense of self, which is not embedded within formal, unmodifiable ideas of subjectivity, but is more apt to be positioned in a space of both political and erotic possibility. In articulating my own version of femme lesbian identity I recognize that this gender and sexuality has a history that has yet to be fully recorded and will often exist unseen by conventional views. I've had to piece together a version of gender and sexuality in preparation for this challenge. While patterns appear through words and images, it is necessary to recognize that the narrative that becomes institutionalized, vertically overshadows landscapes where multiple identities may converge. The gaydar technology that attempts to read a vibe must learn to translate complexity inside and outside of the gender and sexual binary if it is to be an effective method for socially mapping identity, and even then, it must take into consideration the possibility that personal reality can shift, and therefore evade the technology that categorically constructs difference. In addition, this point of interstice that undermines dominant categories demonstrates a consciousness in opposition, which utilizes multiple mechanisms of resistance, and provides for the practitioner, movement and flow between ideological grounds in order to render them even more unstable and problematic.

ISSUES AND QUESTIONS FOR FURTHER CONSIDERATION

1. How can certain gender performance such as that represented by femme identity be both subversive and reinforcing of the cultural expectations of gender?
2. What might be the purposes and limitations of *gaydar* in queer or lesbian social settings?
3. What are the connections between femme identity and transgender identity in terms of both the *gay vibe* and the strategic use of differential consciousness?

REFERENCES

Barthes, R. (1972). *Mythologies.* New York: Hill and Wang.

Butler, J. (1996). Imitation and Gender Insubordination. In A. Garry and M. Pearsall, (Eds.), *Women, Knowledge, and Reality: Explorations in Feminist Philosophy* (2nd ed.). (pp. 371-387). New York: Routledge.

Butler, J. (1999). *Gender Trouble: Feminism and the Subversion of Identity.* New York: Routledge.

Davies, C. B. (1996). Other Tongues: Gender, Language, Sexuality and the Politics of Location. In A. Garry and M. Pearsall, (Eds.), *Women, Knowledge, and Reality: Explorations in Feminist Philosophy* (2nd ed.). (pp. 339-352). New York: Routledge.

Martin, B. (1996). *Femininity Played Straight: The Significance of Being a Lesbian.* New York: Routledge.

Rednour, S. (2000). *The Femme's Guide to the Universe.* Los Angeles: Alyson Publications.

Sandoval, C. (2000). *Methodology of the Oppressed.* Minneapolis: University of Minnesota Press.

Sandoval. C. (2002). Dissident Globalizations, Emancipatory Methods, Social-Erotics. In A. Arnaldo Cruz-Malave and M. F. Manalansan IV (Eds.), *Queer Globalizations: Citizenship and the Afterlife of Colonialism.* (pp. 20-32). New York: New York. University Press.

Gender Please, Without the Gender Police: Rethinking Pain in Archetypal Narratives of Butch, Transgender, and FTM Masculinity

Madelyn Detloff

SUMMARY. Why is it that many of the often-cited narratives about butch, FTM, and/or transgender masculinity happen to be fictions that highlight suffering as a de facto rite of passage for the butch, FTM, or transgendered protagonist? This essay attempts to answer that question, first by outlining the "coherentist assumptions" of lesbian feminism and the forms of gender-policing that cast butch, FTM, and transgendered

Madelyn Detloff is Assistant Professor of English and Women's Studies at Miami University of Ohio. She is currently working on a book titled *The Apocalyptics of Subjection: Trauma, Modernity, Complexity.* Her publications include: "'Father, Don't You See I'm Burning?': Identification and Re-membering in H.D.'s World War II Writing," in *Incest and the Literary Imagination,* University Press of Florida (2002); "'Thinking Peace into Existence': The Spectacle of History in *Between the Acts,*" *Women's Studies* 28 (1999); "Mean Spirits: The Politics of Contempt Between Feminist Generations," *Hypatia* 12.3 (1997); and "Idealized, Debased, and Ordinary: Gender in (Post)Modern Circuits of Desire," *Modern Fiction Studies* 47.4 (2001).

Address correspondence to: Madelyn Detloff, Department of English, Miami University of Ohio, Oxford, OH 45056 (E-mail: detlofmm@muohio.edu).

[Haworth co-indexing entry note]: "Gender Please, Without the Gender Police: Rethinking Pain in Archetypal Narratives of Butch, Transgender, and FTM Masculinity." Detloff, Madelyn. Co-published simultaneously in *Journal of Lesbian Studies* (Harrington Park Press, an imprint of The Haworth Press, Inc.) Vol. 10, No. 1/2, 2006, pp. 87-105; and: *Challenging Lesbian Norms: Intersex, Transgender, Intersectional, and Queer Perspectives* (ed: Angela Pattatucci Aragón) Harrington Park Press, an imprint of The Haworth Press, Inc., 2006, pp. 87-105. Single or multiple copies of this article are available for a fee from The Haworth Document Delivery Service [1-800-HAWORTH, 9:00 a.m. - 5:00 p.m. (EST). E-mail address: docdelivery@ haworthpress.com].

doi:10.1300/J155v10n01_05

subjectivities as "false consciousness." It then analyzes the practice of anchoring butch, FTM, and/or transgender identity claims in pain-filled narratives such as *The Well of Loneliness, Stone Butch Blues,* and *Boys Don't Cry.* While these narratives do important cultural work–exposing the violence heaped upon butches, FTMs, and transgendered guys–it may be time to imagine alternative narratives that are less invested in suffering as a barometer of masculine authenticity. *[Article copies available for a fee from The Haworth Document Delivery Service: 1-800-HAWORTH. E-mail address: <docdelivery@haworthpress.com> Website: <http://www.HaworthPress.com>* © *2006 by The Haworth Press, Inc. All rights reserved.]*

KEYWORDS. Butch, FTM, transgender politics, tragic narratives, lesbian feminism, coherentist assumptions, gender-policing, violence

It seemed to me, and continues to seem, that feminism ought to be careful not to idealize certain expressions of gender that, in turn, produce new forms of hierarchy and exclusion. In particular, I opposed those regimes of truth that stipulated that certain kinds of gendered expressions were found to be false or derivative, and others, true and original. The point was not to prescribe a new gendered way of life that might then serve as a model for readers of the text. Rather, the aim of the text was to open up the field of possibility for gender without dictating which kinds of possibilities ought to be realized.

–Judith Butler[1]

As gender-queer practices and forms continue to emerge, presumably the definitions of "gay," "lesbian," and "transsexual" will not remain static, and we will produce new terms to delineate what they cannot. In the meantime, gender variance, like sexual variance, cannot be relied on to produce a radical and oppositional politics simply by virtue of representing difference. . . . I suggest we think carefully, butches and FTM's alike, about the kinds of men or masculine beings that we become and lay claim to: alternative masculinities, ultimately, will fail to change existing gender hierarchies to the extent to which they fail to be feminist, antiracist, and queer.

–Judith Halberstam[2]

DISAPPEARING ACTS:
COUNTERIDENTIFICATION
AND COHERENTIST ASSUMPTIONS

I still recall my first encounter with Leslie Feinberg's *Stone Butch Blues*, a text that is arguably *The Well of Loneliness* of transgender/ FTM subjectivity . . . or would be so, if *The Well* were not, as Jay Prosser suggests, already in itself a good candidate for *The Well of Loneliness* of transgender/FTM subjectivity.[3] Feinberg's protagonist, Jess Goldberg, painfully laments the loss of butch/femme culture during the heyday of lesbian feminism (roughly the '70s through the '90s) in the long, excruciating epistle that begins hir narrative and provides the occasion for the unfolding of hir subsequent bildungsroman.[4] Addressing hir estranged lover, Theresa, Jess bemoans the gender-policing that drove transgender butches such as hirself out of the safe(r) circles of lesbian community into the lonely and perilous world of passing in homophobic, transphobic, and butchphobic dominant culture. "We thought we'd won the war on liberation when we embraced the word gay," Jess writes:

> Then suddenly there were professors and doctors and lawyers coming out of the woodwork telling us that meetings should be run with Robert's Rules of Order. (Who died and left Robert god?)
> They drove us out, made us feel ashamed of how we looked. They said we were male chauvinist pigs, the enemy. It was women's hearts they broke. We were not hard to send away, we went quietly. (SBB 11)

The bemusing prospect of imagining a lesbian feminist gathering run according to Robert's Rules of Order notwithstanding (pity the poor dyke who would dare make such a patriarchal suggestion to the collective!), Jess's history of butch/FTM disappearance is somewhat overstated, not because it exposes lesbian feminists' accusations of "male identification" for their intolerance of female masculinity, but rather for the surprising power that it attributes to lesbian feminism. While lesbian feminism may have been hegemonic within certain lesbian communities marked, as Esther Newton argues, by "class and race ignorance and antagonisms dressed in new, ideological clothes," its larger influence was nevertheless limited by homophobia and sexism within dominant culture.[5] I somehow missed the hoards of lesbian feminist professors, doctors, and lawyers coming out of the woodwork when I was a baby butch coming out in the mid 1980s, nor do I recall the disappearance of butches and FTMs as being quite as complete as Jess's 20-year exile would suggest.

While it is true that butches and FTMs were being excoriated for "false consciousness" in the pages of scholarly journals and in lesbian feminist publications such as *off our backs*, my sense is that these forms of identity-policing did not force the *disappearance* of butch or FTM subjectivities, but rather shored up the boundaries of a particular historically, culturally and politically situated form of imagined community–lesbian feminism–through counter-identification with the bogey of female masculinity, or masculinity and femininity in general.[6] For example, anthropologist Deborah Goleman Wolf, a self-identified heterosexual woman who spent several years as a participant observer in a lesbian feminist collective in the early 1970s, reports that many of the lesbian feminists she interviewed characterized butch and femme roles as a marker of "old gay" life before the time of women's and gay "liberation."[7] Newton argues that the stigmatization of 'role playing' (butch or femme self-identification) within lesbian feminist movements betrays the class bias of lesbian feminism, which she characterizes as predominantly white and middle class.[8] In lesbian feminist ideology, according to Newton:

> Working-class women, black, brown, and white, gay and straight, had "low consciousness" unless they "cleaned up their act," that is, became more middle class. Unfortunately, these were largely old class putdowns clothed in new political sanctity. . . . In reaction, most working-class women, gay and straight, have shied away from the movement. Working class lesbians stayed in the bars and on softball teams.[9]

In a similar vein, Henry Rubin argues that the ideology of lesbian feminism–namely, a preference for the "woman identified woman" as the "revolutionary subject" of feminism–contributed to the rise of FTM subjectivity as that which was excluded from representation within lesbian feminism: "This revolutionary *excess* consolidated itself as female-to-male transsexualism, giving male-identified individuals another subject position to inhabit."[10]

That lesbian feminism may have imagined itself as *the* women's community, or *the* lesbian community, does not make it so. As Newton argues in her analysis of sociological and anthropological studies of "the lesbian community," quantitative studies have "grossly over-represented the proportion of white, college-educated women in the lesbian 'community'"[11] "Calling a lesbian-feminist group under study 'the [lesbian]community,'" Newton continues, "follows native practice, but is a terrible error from the etic (observer's) point of view."[12] Nevertheless, the rhetorical attacks leveled at butches and FTMs in the name of lesbian feminism–suggesting that female masculinity is a sign of "internalized misogyny," or that butch/femme role playing "honor[ed] the dictates of a warped, male-managed culture"–certainly worked to create yet another normative discourse equating female masculinity or male-identification with moral deviance.[13] Given that Jess might have expected a more accepting attitude on the part of women who were also battling the forces of heteronormativity, it is understandable that sie, who found precious little haven from the violences of heteronormative, transphobic, and sexist dominant culture, would have experienced the ostracism of lesbian feminism as a particularly cruel betrayal–the latest in a long string of assaults on hir right to exist as a transgendered "he-she."

The question to ask at this point is "why?" Why pick on butches and FTMs when, as visibly marked "deviant" bodies, they clearly had less access to social and economic privilege than someone who might enjoy the fragile protections of the closet, or of heterosexual presumption? Why would butch masculinity or FTM masculinity be threatening to lesbian feminism? I would suggest that part of the answer (the part that is not simply transphobic) has to do with the political goals of lesbian feminism. As an identity politics movement, one that attempts to represent (to speak for) lesbians and women, lesbian feminism necessarily relies on what Janet Halley calls "coherentist assumptions" in order to articulate the political interests of the group.[14] Coherentist assumptions, for Halley, do more than *describe* the shared traits of a core constituency; rather, they construct the shared traits that will count as the markers of identity, and distinguish the qualities that will count as 'false consciousness,' or betrayals of the core constituency. That is, coherentist assumptions work to interpellate subjects "from below, from within resistant social movements."[15] Butches and FTMs resist their interpellation as "women" from above–that is, from dominant cultural formations–as well as from below, from the subcultural formation of lesbian feminism. They therefore call into question the coherentist assumptions of lesbian feminism and are cast as betrayers of the "imagined community" lesbian feminism represents.

Such imagined communities rely for their sense of coherency upon border policing, or gender policing in the case of butches and FTMs. This type of border policing can operate reciprocally, when new subcultural formations develop to adapt to the exclusions of interpellation from above and below. As Jacob Hale notes, such is the case in FTM communities, where FTMs or transmen often feel affronted when they are mistakenly assumed to be butch or lesbian. For Hale, it is "so important for some FTMs to distinguish themselves from butches" because coherentist identity assertions rely upon seemingly contiguous (and therefore potentially contagious) forms of gendered subjectivity:

> Identity is always doubly relational (at a minimum). We form and maintain our identities by making continually reiterated identifications *as* members of some category U(s). This is accomplished both positively and negatively by repeated identifications *with* some (not necessarily all) members of U, and by reiterated identifications *as* not-members of some other category T(hem) . . . Some members of U serve as positive identificatory referents, whereas some members of T serve as negative identificatory referents. For many FTMs, lesbians–and especially butches because of their masculinity–serve as primary negative identificatory referents.[16]

Similarly–because I don't want to suggest that counter-identification is a phenomenon exclusive to butches, or FTMs, or lesbian feminists– assimilationist constructions of contemporary lesbian community also imagine contiguous forms of identification negatively.[17] Positive criticisms of the new Showtime series *The L Word,* for example, trumpet "our" apparently long-awaited arrival in the Promised Land of gender normativity, where we are free at last from "our" embarrassing associations with bolo-sporting, Birkenstock-wearing, U-Haul-driving, feminist-professing, frumpy lesbian representations.[18] Kera Bolonik, a reviewer for *New York Magazine,* for example, lauds the show for its glamorous depiction of sexy, financially successful, feminine lesbians: "But after years of living down our dumpy reputation, perhaps it behooves us to put our best, most made-up faces forward, for a change. I mean, how many 'anomalous' dykes does it take to prove to straights and gay boys that not all lesbians wear bolo ties and Birkenstocks?"[19] Just as butches and FTMs were the negative referent for lesbian feminism's coherentist assumptions, lesbian feminists (their Birkenstocks give them away) are the negative referent for today's "new" lesbians, who are, as the review title suggests, "Not your mother's lesbians."[20]

Lost in all of this name-calling is a positive articulation of pleasure, and the non-identitarian affiliations it might engender, for merely accusing others (old-fashioned butches, coupled lesbian moms, lesbian feminist dykes, etc.) of lacking, or insufficiently desiring, pleasure, is not the same thing as affirming it. Because pleasure has been so effectively denigrated as dubious, irresponsible, dangerous, in an era when biopower requires the internalization of disciplining norms, it has become something that must be justified either because of its radical transgressiveness or its containment within loving, heteronormative socially sanctioned relationships. In either case, we might remember Foucault's caution that, "A suspicious mind might wonder if taking so many precautions in order to give the history of sex such an impressive filiation does not bear the traces of the same old prudishness: as if those valorizing correlations were necessary before such a discourse could be formed or accepted."[21]

It is my suspicion that the maligned status of pleasure for its own sake accounts for the prominence of unpleasure and pain in archetypal narratives of butch or FTM masculinity. By "archetypal," I mean a familiar and repeated narrative structure that seems to "make sense" to a culture, and to be compelling enough to re-occur as a dominant mode of telling particular kinds of stories. I am drawing very loosely on Hayden White's understanding of the archetypal emplotments of historical narratives, and his suggestion that narrative *forms* convey meanings that exceed the *contents* of particular narratives.[22] (Unlike White, I don't presume that such archetypal emplotments are transhistorical or transcultural.) While there are narratives of butch or FTM subjectivity that are not angst-filled (for example, Jamison Green's autobiography *Becoming a Visible Man*, the novel *Patience and Sarah*, or the film *The Incredibly True Adventures of Two Girls in Love*), these tend to be less often cited as narratives about the butch or FTM experience than pain-filled narratives such as *The Well, Stone Butch Blues*, or *Boys Don't Cry*.[23] Such narratives do the cultural work of presenting butch, transgender, or FTM subjectivities as valorized, authentic, and deserving of recognition.[24] As Jean Bobby Noble suggests, the protagonists of such narratives (including Mary/Martin, the protagonist of Rose Tremain's *Sacred Country*), share "the belief that masculinity is somehow coterminous with suffering."[25] But why should pain authorize masculine recognition or authenticity? The remainder of my essay seeks to answer that question, and to interrogate its terms.

POLICING GENDER THE OLD FASHIONED WAY

I first included the term "police" in my title as a play on words–suggesting that the identity policing that Feinberg and Newton, among others, describe, also polices pleasure–with the result that we have too much policing and too little pleasing. But the police make more than a semantic appearance in the two most popular representations of transgender or FTM subjectivity in mainstream culture–Kimberly Pierce's 1999 film, *Boys Don't Cry*, and Feinberg's *Stone Butch Blues*. Both of these representations have taken on something of the status of *The Well of Loneliness* for contemporary audiences, and both share *The Well's* penchant for tragic emplotment. Pierce's film differs from the two novels significantly in that it is not a self-authored FTM narrative, but is rather a third-person fictionalized dramatization of the events that lead to the brutal 1994 rape and murder of a female-bodied masculine youth in Falls City, Nebraska. Both Halberstam and Jacob Hale have analyzed the film for what it leaves out of the story–Brandon's undecided "borderland" status between identity categories which Brandon never explicitly claimed for himself (transgender, FTM, transsexual, stone butch . . .), for the narrative's depiction of rural culture, for the absence of mention of race and racism as part motivation for the murders. Therefore I won't rehash the film's limitations as an accurate representation of historical events here.[26] Rather, I am interested in how the film mobilizes the generic elements of tragedy in order to present a portrait of FTM subjectivity that *seems* coherent precisely because it successfully deploys the already available tragic plot structures of *The Well* and *Stone Butch Blues*.[27]

While it is important to identify and analyze the cultural work done by the tragic plot structures of these popular FTM narratives, I want to be clear that my intention is not to replicate lesbian feminism's well-known condemnation of *The Well* for its tragic structure–"the 'butch,' the tears, the despair of it all" in the words of Wiesen Cook.[28] Rather, I would like to focus on what that tragic literary form does, especially since, as Elizabeth Bruss argued almost 30 years ago (ironically during the same time period that lesbian feminists were excoriating butches, femmes and transsexuals for false consciousness), "A literary institution must reflect and give focus to some consistent need and sense of possibility in the community it serves, but at the same time, a genre helps to define what is possible and to specify the appropriate means for meeting an expressive need."[29] Given that, as Bruss argues, "conceptions of individual identity are articulated, extended and developed" through such institutions–then the project of enabling less lethal imaginations of butch and FTM identities might be furthered by analyzing the forms of narratives that foreclose possibilities for living on, beyond tragic or melancholic modes of identification/identity formation.

As Judith Butler argues much more recently, "thinking of the possible" is a vital and necessary political/ethical act, one that is foreclosed by gender regulations that posit certain subjectivities (and bodies) such as FTM or intersexed, as unthinkable or unimaginable:

> The point is not to prescribe new gender norms, as if one were under an obligation to supply a measure, gauge, or norm for the adjudication of competing gender presentations. The normative aspiration at work here has to do with the ability to live and breathe and move and would no doubt belong somewhere in what is called a philosophy of freedom. The thought of a possible life is only an indulgence for those who already know themselves to be possible. For those who are still looking to become possible, possibility is necessity.[30]

The appearance of the police in *Stone Butch Blues* and *Boys Don't Cry* intensifies the stakes of the kind of gender regulation that Butler analyzes in *Undoing Gender*, for their presence is both symbolic–upholding the cultural, psychological, and juridical violences of gender normativity–and literal. They at once stand for disciplinary power (of the sort that D.A. Miller describes in *The Novel and the Police*), and a repressive, thuggish power that exercises itself through rape, battery, and psychological violence.[31] It is crucial to acknowledge the latter form of gender-norm enforcement (through what Louis Althusser calls Repressive State Apparatuses) as the brutal corollary to the ideological gender policing that produces norms for proper and improper, authentic and inauthentic gender performances.[32] These two forms of gender policing (the repressive and the ideological) are not polar opposites, but rather are entangled in a complicated relationship to wounding. While the police, for many marginalized communities, stand as dominant culture's sanctioned and sanctioning wounding agents, the psychological wounds of marginalization also function as motivation for "grassroots" gender policing among marginalized communities.[33] And, as Wendy Brown reminds us, the maintenance of politicized identity relies upon a reiteration of the condition of woundedness as guarantor of its future existence.[34]

This predicament of "wounded attachment," in Brown's words, is not something to be taken lightly in the face of a dominant culture that would rather that gender deviants such as butches and FTMs not exist. However, there is something deeply troubling when the maimed and wounded butch or FTM body becomes the ultimate cultural anchor for the reality of butch or FTM suffering and existence. As Elaine Scarry notes, in an unfortunately not-outmoded analysis of the structure of war, the body serves as "the ultimate source of substantiation" for cultural constructs that have been "derealized."[35] Substantiation, for Scarry, involves:

> the extraction of the physical basis of reality from its dark hiding place in the body out into the light of day, the making available of the precious ore of confirmation, the interior content of human bodies, lungs, arteries, blood, brains, the mother lode that will eventually be reconnected to the winning issue, to which it will lend its radical substance, its compelling, heartsickening reality.[36]

In tragic representations of butch and FTM gender performance, the wounded body similarly substantiates the as yet unanchored identities of the butch and FTM protagonists. For example, in *The Well of Loneliness*, as Jodie Medd demonstrates, Stephen's war wound serves both as literal proof of her valor, and as a proxy for her wounded status as a gender deviant. "The actual war wound and Stephen's figurative 'war of existence' powerfully condense in the scar she receives from a splinter of a shell while she serves as an ambulance driver at the front," Medd explains. "Stephen's fixation upon bearing the 'mark of Cain' insists that we read her scar as the physical iconography of her invert suffering."[37]

Sally Munt suggests that the novel "reads inversion through the lens of Christian martyrdom and agency," and therefore operates as something of a "well of shame."[38] Clearly, *The Well* connects·Stephen's developing masculine identification and her "invert" desire for a feminine love object to her capacity for suffering. In a scene that explicitly foreshadows the development of Stephen's adult "invert" subjectivity as a nexus of suffering, sacrifice, transgender identification, and desire for the feminine female, the young Stephen's inversion is revealed (to the readers, if not to the young and suffering Stephen) through her masochistic infatuation for her family's housemaid, Collins. Forcing herself to kneel on the floor of her nursery until her knees were "particularly scarified," Stephen imagines herself a condensation of Lord Nelson (I'm in the middle of the Battle of Trafalgar–I've got shots in my knees!") and Christ: "I would like very much to be a Saviour to Collins" she prays, "I love her, and I want to be hurt like you were" (*Well*, 21-22). Moreover, her desire for Collins and for suffering are depicted as one and the same–"it was really rather fine to be suffering–it certainly seemed to make Stephen feel that she owned her by right of this diligent pain" (*Well*, 23).

It is not the masochistic aim that is troubling in this formation, but the suggestion that it forms the kernel of Stephen's "invert" subjectivity, a kernel that will unfurl inevitably according to its preprogrammed course, until the heroic, scarified, sacrificial, desiring, and imploring masculine invert fully emerges at the end of the novel.[39] In *Well*, and its late 20th century progeny (*Stone Butch Blues* and *Boys Don't Cry*), the butch or FTM characters seem compelled to make decisions that are against their best interests, endangering their safety and/or psychological well-being. (In *Well*, the sacrificial violence is mostly psychological–a telling commentary on the protective benefits of class privilege.) Moreover, the tragic structure of all three of these narratives seems to imply that female masculinity is something of an inescapable tragedy in itself. At the very least, such tragic emplotments mobilize what Michael André Bernstein calls in another context the "logic of inevitability"–the inevitability of butch and FTM suffering–and this is a logic I would like to contest.[40]

In *Boys Don't Cry*, for example, Pierce's tragic emplotment of the events surrounding Brandon's murder coalesce with a narrative of development–Brandon's attempt to construct a legible masculine identity for himself. He does this through cross-dressing, a practice the film attends to in detail, and participating in various self-destructive forms of male bonding, such as alcohol abuse, bar fighting, reckless driving and "bumper skiing." When Lana, Brandon's love object in the film, asks him why he would let someone tie him to the back of a pickup truck and drag him through the mud, he replies, "I just thought that's what guys did around here." Furthermore, the film is self-announcing about the connection it makes between masculinity and masochism, especially in a scene where, cowboy-like, Brandon and Tom Nissen sit around a campfire while Tom mentors Brandon in masculinity by telling him how he cuts himself in order to survive the psychological pressures of prison. "Real" men–macho men–can cut deeper and still master their pain. In a moment of dramatic irony, Brandon betrays to the audience, but not to Tom, his "true" identity when he refuses Tom's invitation to participate in the ritual of self-mutilation, saying, "I guess compared to you I am a pussy." These words haunt the film, and foreshadow its inevitable violence.

Despite Brandon's efforts at gendering, the film implies that it is the law that will ultimately determine Brandon's gender identity. Indeed, Brandon's *hamartia*–his error in judgment–seems to be his inability to submit to the law's categorization of him as female. (The tragic logic of the film suggests–rather transphobically–that if only he had answered to the court's summoning of "Teena Brandon" he, or rather "she," would be "safely" in jail and not free to risk his life by trying to pass in Falls City, Nebraska.) Indeed, it is because of his legal troubles that Brandon's "true" sex is disclosed to his eventual rapists and murderers, who further enforce the "reality" of the law's juridical assignment of gender identity by raping him, thus, in their twisted logic, "making" him a woman–literally, in a recycling of Brandon's misogynist epithet, just a "pussy." The police interrogation that follows the rape further solidifies this collusion between the police and the perpetrators. The sheriff asks Brandon if the perpetrators who de-pantsed him were men, and then asks why didn't they attempt to penetrate him when they took down his pants and "found out [he] was a woman." Continuing the interrogation, the sheriff then asks Brandon if he is a virgin, presuming that FTM or butch sex with women is not "real" sex, while being violently pene-trated in a rape is. Ultimately the sheriff's questions force Brandon to invoke his violated body ("my vagina") in ways that legitimate his violators' assignation of gender identity.

This police collusion with the rapists' "enforcement" of gender iden-tity is shown in the film to be problematic, which is why it is even more puzzling to find the film colluding in less overt ways with the rapists' gender assignation in the highly improbable sex scene which takes place after the rape. In this scene, suddenly and inexplicably, Brandon allows Lana to unbind his breasts and presumably make love to him, presumably as a lesbian woman. Halberstam analyzes this scene in de-tail, noting that "the scene implies that the rape has made Brandon a woman in a way that his brutal exposure earlier in the bathroom and his intimate sex scenes with Lana could not."[41] If the rape, in the logic of the film, does make Brandon a woman, then the narrative disturbingly does in Brandon the FTM before John Lotter and Tom Nissen finish the job.[42]

While Pierce's film ultimately (and problematically) consolidates gender "realness" in its "dark hiding place in the body," Feinberg's *Stone Butch Blues* is much more committed to the viability of a transgendered subjectivity. As in *Boys Don't Cry*, however, the borders of this subjectivity are also enforced by violent policing, quite literally, by the violent police. In the novel, the rapist police are apparently endowed with the ability to distinguish the real butches from the fakes, the "Saturday night butches." Jess, the protagonist of the novel, recalls:

> I remember the busts in the bars in Canada. Packed in the police vans, all the Saturday-night butches giggled and tried to fluff up their hair and switch clothing so they could get thrown in the tank with the femme women–said it would be like "dyin' and goin' to heaven". The law said we had to be wearing three pieces of women's clothing.
>
> We never switched our clothing. Neither did our drag queen sisters. We needed our sleeves rolled up, our hair slicked back, in order to live through it. (SBB, 8)

The "Saturday night butches" are, according to the novel, rarely taken into to police custody, or are taken in and then booked and released, while the real butches are beaten, raped, and otherwise tortured. This is an important articulation of police brutality against FTMs and butches, but it's hard to imagine that such thuggish police would care to make distinctions between "authentic" performances of female masculinity and "inauthentic" ones. Indeed, Feinberg seems to grant the police finer powers of gender discernment than her femme lover:

> "You're a woman!" Theresa shouted at breakfast. She pushed her plate away. Her part-time temp work had put that meal on the table.
>
> "No I'm not," I yelled back at her. "I'm a he-she. That's different."
>
> Theresa slapped the table in anger. "That's a terrible word. They call you that to hurt you."
>
> I leaned forward. "But I've listened. They don't call the Saturday-night butches he-shes. It means something. It's a way we're different. It doesn't just mean we're . . . lesbians." (SBB, 147-48)

This is an important articulation of the difference between transgendered masculinity and non-transgendered butch lesbianism–parsing the often-conflated categories of sex, sexuality, and gender. To suggest that one's susceptibility to abuse from the representatives of dominant culture and its laws become the arbiter of that difference, however, replicates the logic of wounded attachment, one that, as Brown notes, forecloses the possibility of imagining an identity that is not dependent for its definition on the persistence of injury.[43] Should the police–cast in Feinberg's narrative as the violent, phobic, repressive agents of dominant culture–ultimately be the forces who distinguish FTMs from butches from other lesbians, setting up a hierarchy of more or less real or more or less fake masculinities? This form of gender policing seems to be a narrative practice that limits, rather than opens up, possibilities for viable gender identities and performances.

What does it mean that the current archetypal narratives of butch, FTM, or transgendered masculinity tend to produce a sense of "authentic" identity through genres that present suffering as inevitable and persistent? Asking this question does not dismiss the very real physical and psychological violence heaped upon butches, transgendered guys, and FTMs under the wink and nod of an intolerant and phobic dominant culture. Nor is it intended to contribute to yet another psychological "profile" of butch, trans, or FTM folks, surmising that they simply wish violence upon themselves and then get it from a dominant culture only too willing to oblige. Instead, we might rather recognize violence as an aspect of the cultural conditions under which butches, FTMs, and transgendered guys live, rather than as an element constitutive of their masculinity. It is important to understand this violence as part of history, rather than as an element of ontology, so that we might imagine FTM, butch, and transgendered subjectivity in a future that is not haunted by such violence. In the process, we might even imagine ways to perform other (e.g., biologically-born male) masculinities in ways that do not conflate vulnerability with shame and femininity.

As Bernstein suggests, "In the domain of history, unlike in the world 'seen with the eyes of the [tragic] genre,' there are always multiple paths and sideshadows, always moment-by-moment events, each of which is potentially significant in determining an individual's life, and each of which is a conjunction, unplottable and unpredictable in advance of its occurrence, of specific choices and accidents."[44] Imagining an unpredictable, open future for the non-tragic formation of FTM, butch, and/or transgendered subjectivities is more than a matter of providing accurate descriptions of gendered realities. It requires imagining into existence gendered possibilities that are both livable and respected. As Teresa de Lauretis argues, "public fantasies" such as novels and films "contribute to the shaping of the social imaginary," and "provide material and scripts, or forms of content and expression, to the subjective activity of fantasizing."[45] "Public fantasies" make certain forms of desire and certain modes of embodiment thinkable, imaginable, and therefore available for the creation of culturally intelligible sexed and gendered subjects. It's not too much to ask, therefore, that we cultivate public fantasies of FTM, butch, and transgendered masculinities that are life-affirming and pleasurable, as well as attendant to the realities of phobic violence. In other words, it is time to imagine a future for non-normative masculinities that are more filled with gender pleasing than gender policing.

ISSUES AND QUESTIONS FOR FURTHER CONSIDERATION

1. How do we define "the lesbian community"? How might we foster descriptive understandings of subcultures or communities without turning our descriptions into prescriptive norms?
2. The phrase "the real world" is often used as a euphemism for a tough life, or the "school of hard knocks." Are there ways to acknowledge the difficult experiences faced by women, lesbians, FTMs, butches (and others marked by unequal gender and sexuality norms) without overvaluing pain as the measure of the 'reality' of one's experience?
3. What kinds of "public fantasies" or narrative forms might work as alternatives to the pain-filled archetypal narratives that seem predominant in stories about female or FTM masculinity? What sort of political/cultural work might these alternative narratives do?

AUTHOR NOTE

Many thanks to Angela Pattatucci Aragón and the anonymous reviewers from *JLS* for their suggestions for improving this essay. My gratitude as well to Robyn Wiegman for her as-always generous encouragement and her incisive comments on an earlier form of this essay, as well as to Judith Halberstam for graciously sharing a manuscript copy of her conference presentation on "Transgender Feminism and the Evolution of the Clownfish" with me. Thanks to Karen Mitchell, who was an indispensable and indefatigable research assistant for this project during crunch time, and to Susan Pelle, who provided impromptu research assistance at the 11th hour. Of course, any errors, omissions, or misprisions in this article are my own.

NOTES

1. Judith Butler, *Gender Trouble*, 10th Anniversary Ed. (New York: Routledge,1999), viii.

2. Judith Halberstam, *Female Masculinity*. (Durham: Duke University Press, 1998), 173

3. There are good reasons, as Prosser argues, for reading *The Well of Loneliness* as an early transsexual narrative. Stephen Gordon, the protagonist, identifies as an "invert,"

which in sexological discourse of the early 20th century would have referred to a male psyche 'trapped' in a female body. However, reading the novel as a transsexual narrative does not necessarily preclude reading it as the story of a masculine female. Jean Bobby Noble, for example, suggests that "*The Well* is a discursive event wherein the histories of white masculinity as male-embodied, female masculinity as butch- and lesbian-embodied, and trans-sexual masculinity as meta-embodied (and failed embodiment) all overlap" (49). We can't know whether Hall would have identified Stephen differently if different categories (FTM, transsexual, transman) were available to her. Our very inability to disarticulate the categories retroactively is perhaps a motivation for transmen such as Prosser to carefully differentiate categories in the present day. By "FTM," I mean a person who lives his life as a male, despite being assigned to the category of "female" at birth. By "transgender," I mean a person who does not live his/her life conforming to the sex category "male" or "female" assigned at birth. For the sake of this argument, I will follow Halberstam's assumption that there are a wide variety of ways to embody a transgender subjectivity. Jay Prosser, "'Some Primitive Thing Considered in a Turbulent Age of Transition': The Transsexual Emerging from *The Well*," in *Palatable Poison: Critical Perspectives on the Well of Loneliness*, ed. Laura Doan and Jay Prosser (New York: Columbia University Press, 2001), 129-144; Leslie Feinberg, *Stone Butch Blues* (Ithaca, NY: Firebrand Books, 1993); Radclyffe Hall, *The Well of Loneliness* (New York: Anchor Books, 1990); Judith Halberstam, *Female Masculinity* (Durham: Duke University Press, 1998); Jean Bobby Noble, *Masculinities Without Men?* (Vancouver: UBC Press, 2004). Subsequent references to *The Well of Loneliness* will be abbreviated as *Well* and cited parenthetically in the text. Subsequent references to *Stone Butch Blues* will be abbreviated as *SBB* and cited parenthetically in the text.

4. I respect Feinberg's desire to use the gender-neutral pronouns "sie" (third person nominative) and "hir" (third person possessive), and will therefore use the pronouns to refer to Jess, Feinberg's protagonist throughout this essay. For a discussion of hir pronoun preferences, see Feinberg's "We Are All Works in Progress," in *TransLiberation* (Boston: Beacon Press, 1998), 1-13.

5. Esther Newton, *Margaret Mead Made Me Gay: Personal Essays, Public Ideas* (Durham: Duke University Press, 2000), 161.

6. Many scholars of femme identities have pointed out lesbian feminism's counter identification with femme femininity as well as female masculinity. See, for example, Biddy Martin, "Sexualities Without Genders and Other Queer Utopias," *Diacritics* 24 no. 2-3 (1994):104-21; and Laura Harris and Liz Crocker, "An Introduction to Sustaining Femme Gender," in *Femme: Feminists, Lesbians, and Bad Girls*, ed. Laura Harris and Elizabeth Crocker (New York: Routledge, 1997), 1-12. For lesbian feminist disavowals of female masculinity, see, for example, Blanche Wiesen Cook, "'Women Alone Stir My Imagination': Lesbianism and Cultural Tradition," *Signs* 4 no. 4 (1979):718-739; Deborah Goleman Wolf, *The Lesbian Community* (Berkeley: University of California Press, 1979); and Jennie Ruby, "Is the Lesbian Future Feminist?" *off our backs*, October 1996, 22. The term "imagined community" comes from Benedict Anderson, *Imagined Communities: Reflections on the Origin and Spread of Nationalism* (New York: Verso, 1991).

7. Wolf, 40-48.

8. Newton, 156.

9. *Ibid.*, 173.

10. Henry Rubin, *Self-Made Men: Identity and Embodiment Among Transsexual Men*, (Nashville: Vanderbilt University Press, 2003), 64-65. (Rubin's emphasis.)

11. Newton, 156.

12. *Ibid.*, 157. (Newton's insertion).

13. Wiesen Cook characterizes butch/femme relationships as antiegalitarian in "Women Alone Stir My Imagination," 720. While I had originally thought that lesbian feminism's quarrel with butch and FTM subjectivities was largely a historical phenomenon, it is apparently kept alive through media as the *Questioning Trans Politics* Website (*http://www.questioningtransgender.org*), and *off our backs*. (Thanks to the *JLS* anonymous reviewer for calling my attention to the Website.) For example, contemporary lesbian feminist Jennie Ruby describes her own female masculinity as "internalized misogyny" that she overcame through lesbian feminist consciousness in "Identity = Politics: A Personal Story," *off our backs*, April 2000, 9.

14. Janet Halley, "'Like Race' Arguments," in *What's Left of Theory: New Work on the Politics of Literary Theory*, ed. Judith Butler, John Guillory, and Kendall Thomas (New York: Routledge, 2000), 41. Halley draws on the theories of Louis Althusser to nuance the process of "interpellation" or subject-making, suggesting that it is not only "ideological state apparatuses" (dominant cultural institutions such as schools, churches, [normative] families, medical/psychiatric establishments, etc.) that "hail" or call subjects into social recognition, but also subcultural communities which may enforce norms of "acceptable" subjecthood. On "interpellation," see Louis Althusser, "Ideology and Ideological State Apparatuses," in *Lenin and Philosophy and Other Essays*, trans. Ben Brewster (New York: Monthly Review Press, 1971).

15. Halley, 44.

16. C. Jacob Hale, "Consuming the Living, Dis(re)Membering the Dead in the Butch/FTM Borderlands," *GLQ* 4 no. 2 (1998), 330.

17. I use the term "counteridentification," a way to stake an identity position in opposition to an Other ("I am I because I am *not* you"), rather than the term "disidentification," which, as José Esteban Muñoz explains, is a method of "recycling" a "toxic" identification with an identity that is considered abject. José Esteban Muñoz, *Disidentifications: Queers of Color and the Performance of Politics* (Minneapolis: University of Minnesota Press, 1999), 11-12. For example, a femme might disidentify with hegemonic femininity by reiterating its performance in a way that exposes or undermines hegemonic femininity's 'naturalness.' An "androgynous" lesbian feminist might, on the other hand, counteridentify with feminine subjects (hegemonic or femme) entirely, by identifying as not-feminine and similarly counteridentify with not-masculine masculine subjects (bio-males, butches, FTMs).

18. *The L Word*, dir. Rose Troche, prod. Eileen Chaiken (USA: Showtime Networks Inc. USA, 2004-2005). Halberstam critiques the series' lack of butch representation in "The I Word," *Girlfriends*, Feb. 2004, 18.

19. Bolonik, Kera. "Not Your Mother's Lesbians," *New York Magazine*, 12 Jan. 2004. (Available at New York Metro.com *http://newyorkmetro.com/nymetro/news/features/n_9708/.*)

20. Halberstam analyzes the familial dynamics and generational rhetoric implicit in "Not Your Mother's Lesbians" as well as contemporary articles on "boi" culture, asking "What are we to make of these new forms of trans and lesbian culture? Why is feminism posited as both an embarrassing mother who must be pushed aside and a humorless butch aunt who stands in the way of the pursuit of pleasure? Are there other models of generation, temporality, and politics available to queer culture and feminism?" Judith Halberstam, "Transgender Feminism and the Evolution of the Clown-

fish," Paper presented at the Back to the Future: Generations of Feminism Conference, The University of Chicago, February 28 2004, 1.

21. Michel Foucault, *The History of Sexuality, Vol. 1.*, Trans. Robert Hurley (New York: Vintage Books, 1990), 6.

22. Hayden White, "The Historical Text as Literary Artifact," in *Critical Theory Since 1965*, ed. Hazard Adams and Leroy Searle (Tallahassee: Florida State University Press, 1989). White draws on the archetypal criticism of Northrop Frye, *The Anatomy of Criticism* (Princeton: Princeton University Press, 1957).

23. Jamison Green, *Becoming a Visible Man* (Nashville: Vanderbilt University Press, 2004); Isabelle Miller, *Patience and Sarah* (New York: Fawcett Crest, [1969] 1973); Maria Maggenti, *The Incredibly True Adventure of Two Girls in Love*, dir. Maria Maggenti (Fine Line Features,1995).

24. Jay Prosser, for example, reads *The Well of Loneliness* and *Stone Butch Blues* as transsexual and transgendered narratives that "attest to the valences of cultural belonging that the categories of man and woman still carry in our world: what I term 'gendered realness.'" Jay Prosser, *Second Skins: The Body Narratives of Transsexuality* (New York: Columbia University Press, 1998), 11. Jean Bobby Noble reads *The Well, Stone Butch Blues, Boys Don't Cry*, and Rose Tremain's *Sacred Country* as exemplary narratives of female and/or trans-gendered or trans-sexed masculinities. Jean Bobby Noble, *Masculinities Without Men? Female Masculinity in Twentieth-Century Fictions* (Vancouver: University of British Columbia Press, 2004), xii. All of these narratives highlight suffering, and in three out of four (*The Well* excepted), physical violence is a significant plot element. Kimberly Pierce and Andy Bienen. *Boys Don't Cry*, dir. Kimberly Pierce (Fox Searchlight Pictures: 1999). Rose Tremain, *Sacred Country* (New York: Washington Square Press, 1995).

25. Noble, 116.

26. Hale, "Consuming the Living;" Halberstam, "The Brandon Teena Archive" in *Queer Studies: An Interdisciplinary Reader*, ed. Robert Corber and Stephen Valocci (London: Blackwell, 2002), 159-169.

27. One could argue, as Noble does, that *Stone Butch Blues* ends more hopefully, with Jess's coming to consciousness as a transgender and labor activist, but if we read the temporal ending of the narrative as the beginning of the novel–Jess's long and undeliverable letter to her former lover, Theresa–the plot structure becomes much more like that of *The Well*, with the writing of an impassioned plaint the primary form of consolation for the suffering outcast (Noble, 140-41).

28. Wiesen Cook, 718. Catharine Stimpson, in a less condemning tone, analyzes *The Well* for its five-part plot structure, with "each of the novel's five sections (or acts) end[ing] unhappily," and compares *The Well* to the more "hopeful" lesbian novels of the 1970s. Catharine Stimpson, "Zero Degree Deviancy: The Lesbian Novel in English," in *Writing and Sexual Difference*, ed. Elizabeth Abel (Chicago: The University of Chicago Press, 1982), 249.

29. Elizabeth W. Bruss, *Autobiographical Acts: The Changing Situation of a Literary Genre* (Baltimore: Johns Hopkins University Press, 1976), 5.

30. Judith Butler, *Undoing Gender* (New York: Routledge, 2004), 31.

31. D.A. Miller, *The Novel and the Police* (Berkeley: University of California Press, 1988).

32. Althusser, "Ideology and Ideological State Apparatuses."

33. Evelynn Hammonds, for example, suggests that the ideological policing of working class African American women by "some middle-class black women" was a

response to the racist stereotypes of the "oversexualized" black woman used "as a justification for the rape, lynching, and other abuses of black women by whites." Evelynn Hammonds, "Black (W)holes and the Geometry of Black Female Sexuality" in *Feminism Meets Queer Theory*, ed. Elizabeth Weed and Naomi Schor (Bloomington: Indiana University Press, 1997), 142-143. A more disturbing example of this type of gender policing is Janice Raymond's invocation of (birth assigned) women's susceptibility to rape as a justification for her exclusionary (indeed hateful) remarks about transsexual mtfs. Sandy Stone critiques Raymond's dubious claim that "All transsexuals rape women's bodies" in "The Empire Strikes Back: A Posttransexual Manifesto," in *Body Guards*, ed. Kristina Straub and Julia Epstein (New York: Routledge, 1991).

34. Wendy Brown, "Wounded Attachments," in *States of Injury: Power and Freedom in Late Modernity* (Princeton, N.J.: Princeton University Press, 1995).

35. Elaine Scarry, *The Body in Pain: The Making and Unmaking of the World* (New York: Oxford University Press, 1985), 137. Noble cites Scarry's work in a different context, suggesting that Mary/[Martin](the protagonist in *Sacred Country*), and Jess, the protagonist in *Stone Butch Blues*, are "subjects of a kind of domestic daily torture" (Noble, 116).

36. Scarry, 137.

37. Jodie Medd, "War Wounds: The Nation, Shell Shock, and Psychoanalysis in *The Well of Loneliness*," in *Palatable Poison: Critical Perspectives on the Well of Loneliness*, ed. Laura Doan and Jay Prosser (New York: Columbia University Press, 2001), 242.

38. Munt, Sally. "The Well of Shame," in *Palatable Poison: Critical Perspectives on the Well of Loneliness*, ed. Laura Doan and Jay Prosser (New York: Columbia University Press, 2001), 200.

39. Noble discusses Hall's use of masochistic tropes of Christian martyrdom in *Masculinities Without Men?*, 62-67.

40. Michael André Bernstein, *Foregone Conclusions: Against Apocalyptic History* (Berkeley: University of California Press, 1994).

41. Judith Halberstam, *In a Queer Time and Place: Transgender Bodies, Subcultural Lives* (New York: New York University Press, 2005), 90.

42. My analysis here is not meant to suggest that transmen and FTMs would necessarily shun sex that involved caressing, touching breasts, penetration by a partner, etc., but rather that within the context of the film, it seems fairly clear that Lana is making love to Brandon *as a woman*. The scene further suggests that the rape has somehow made his sexual role more "flexible," as if the rape has been sexually liberating to him. Certainly many survivors of sexual violence have maintained active, pleasurable, adventuresome sexual lives, but within the film's narrative logic, the timing of Brandon's sudden change in sexual role seems suspect.

43. Brown, 73.

44. Bernstein, 12.

45. Teresa De Lauretis, "The Stubborn Drive," *Critical Inquiry* 24 (Summer 1998), 866.

Household Remedies:
New Narratives of Queer Containment in the Television Movie

Cait Keegan

SUMMARY. What constitutes lesbian identity and who gets to define and/or inhabit such an identity in this postmodern and mediated world? This article addresses how the structure of televisual discourse restricts and streamlines "lesbian" representations in television movies. The supposed "progress" of appearing in the virtual public spaces of television and print media may fulfill the queer impulse for visibility in opposition to cultural silence, but it may also come at the price of a depoliticization of queer life and erotic resistance. Taking notice of which deployments of "queerness" are created and supported by text of the television movie, this article seeks insight into how the queer body and queer identity are being hegemonically reconstructed for consumption by this media form.

Cait Keegan is completing her PhD in American Studies at the State University of New York at Buffalo, where she also teaches courses in critical white studies. She is an actor, digital filmmaker, and pop culture addict whose scholarly work is specialized in queer sexualities, political theory, and cultural studies. Originally from rural Pennsylvania, Keegan possesses a continuing interest in how the architectures and landscapes of public social bodies affect the political articulation of desire.

Address correspondence to: Cait Keegan, Department of American Studies, 1010 Clemens Hall, State University of New York at Buffalo, Buffalo, NY 14260-4630 (E-mail: greenestgirl@hotmail.com).

[Haworth co-indexing entry note]: "Household Remedies: New Narratives of Queer Containment in the Television Movie." Keegan, Cait. Co-published simultaneously in *Journal of Lesbian Studies* (Harrington Park Press, an imprint of The Haworth Press, Inc.) Vol. 10, No. 1/2, 2006, pp. 107-123; and: *Challenging Lesbian Norms: Intersex, Transgender, Intersectional, and Queer Perspectives* (ed: Angela Pattatucci Aragón) Harrington Park Press, an imprint of The Haworth Press, Inc., 2006, pp. 107-123. Single or multiple copies of this article are available for a fee from The Haworth Document Delivery Service [1-800-HAWORTH, 9:00 a.m. - 5:00 p.m. (EST). E-mail address: docdelivery@haworthpress.com].

Available online at http://www.haworthpress.com/web/JLS

doi:10.1300/J155v10n01_06

KEYWORDS. Television movies, lesbian identity, queer movies, lesbian communities, *Serving in Silence*, Margarethe Cammermeyer, If *These Walls Could Talk Two*, *The Truth About Jane*, *Normal*–the movie

When it comes to lesbians . . . many people have trouble seeing what's in front of them. The lesbian remains a kind of "ghost effect" in the cinema world of modern life: elusive, vaporous, difficult to spot–even when she is there, in plain view, mortal and magnificent at the center of the screen. . . . What we never expect is precisely this: to find her in the midst of things, as familiar and crucial as an old friend, as solid and sexy as the proverbial right-hand man, as intelligent and human and funny and real as Garbo.

　　　　　　　　　　　–Terry Castle, *The Apparitional Lesbian*

There are no lesbians here.

　　　　　　　–Katie King, "There Are No Lesbians Here:
　　　　Lesbianisms, Feminisms, and Global Gay Formations"

There are an awful lot of queers on TV lately. In fact, on today's television, it appears as if our moment in the sun has finally arrived: we queers are overwhelmingly white, male, wealthy, thin, young, fashionable, and sero-negative. We live in expensive urban lofts and spend our free time whipping straights into shape with our catty relationship advice and snappy sense of style. On TV, queers are everything America desires of its citizenry, minus–of course–the annoying problem of the sex we have. Queers on television are already assimilated, already absorbed into the societal framework that popular media assumes for its viewing audience. If television is a community of the imagination, it is one that actively contains and then incorporates the concept of what "queer" means. Examining the processes by which spaces in the televisual discourse both permit and censure queer representations, this article focuses on television movies as sites of cultural meaning and contestation–sites where queerness enters and is transformed by modes of heterosexual containment. The manner in which queerness is "solved" by these narratives reveals a great deal about how our heteronormative society learns to absorb and incorporate sexual difference(s) into its essentially conservative patterns of consumptive community life, as well as what forms of difference remain impossible to absorb and are therefore excluded. In mainstream media texts such as TV movies, this "fixing" of queer difference most often involves a depoliticizing process that reduces queerness to a capitalistic mimicry of heterosexual romantic love–one that no longer threatens the established social fabric with a deconstructive erotic power. Through their recent engagement with specifically lesbian images, television movies function as a case study of the larger society's struggle to incorporate queer people in a manner that does not threaten the assumed cultural meaning of family, home, marriage, economy, and nation. Rather than a flat victory, the appearance of lesbian characters in these movies signals a bargain struck with the forces of political and representational power–a bargain that trades in the erotic edge of queer difference for a set of more mainstream, "acceptable" televisual narratives about the role of queerness in American society.

Television movies constitute a popular media genre that has at times operated centrally in the debate over what is constituted as socially relevant. Without an examination of the generic qualities of television movies, a discussion of the shifts in formula instituted by the introduction of queerness would be next to impossible. Elayne Rapping (1992) makes several extremely significant points about the inherently capitalist and heteronormative structure of made-for-television movies, the first being that the TV movie "redefines the term *family* as much as necessary in order to maintain the myth that all personal problems in a capitalist society may be resolved by individuals who view themselves, essentially, as family members" (Rapping, 1992, p. xi). Telefeatures are particularly focused on the family as a unit of social integration that is capable of absorbing and overcoming the obstacles of discord and political unrest. This is symbolized by their basic structure, which typically depicts a woman-centered family struggling under the fire of controversial social issues. As texts (and products) intimately involved in the support of capitalism, TV movies seek to protect the romantic ideal of the nuclear family that capitalist consumerism depends upon, thereby encouraging participation in the "family values" of shopping and spending. Because they are focused on the integrative patterns of romanticized family life in America, TV movies are not designed to address the public space of the male world, which traditionally advocates competition and rational public debate. Instead, they are directed into the private and feminized space of the home, and are intended for a female audience. The reasons for this address are economic as well as sociocultural. Television has historically been directed toward women as a commercial discourse aimed at tapping the expendable income of housewives. However, TV movies are also particularly concerned with women's issues because they surround and attempt to alleviate a pivotal cultural sticking point:

> In every category of telefeatures dealing with specifically female issues there is a thematic and a structural constant. Thematically, there is the issue of defining the concept of "woman" in an age when female identity is most definitely in social flux and therefore a matter of contested terrain. (Rapping, 1992, p. 96)

It is important to note that the women addressed by TV movies are hailed in a manner that already implies a resolution of the "problem of woman" in patriarchal society. By limiting their direct address to women who play the roles of wives and mothers, traditional TV movies attempt–in a less than always successful manner–to confine the definition of "woman" to a heterosexualized position already embedded in the discourses of patriarchy and capitalism. The audience of TV movies is typically assumed to live an existence that reflects the experience of female characters *within* the televisual discourse. Rapping points out that most TV movies represent women in a very traditional sense, as predominantly wives and mothers, and that "sexually unconventional women are invariably fated to either reform or to come to a bad end" (Rapping, 1992, p. 97).

Additionally, Rapping discusses the limitations that TV movies run up against as realist texts, a critique that is expanded into a wider comment on the conservative nature of television itself, which naturalizes the oppressive conditions of history through the "sham" of bourgeois realist narratives. Realism, which participates in the assumption that all experience is transparent and obvious to the naked (i.e., heterosexual and male) eye, is hegemonically situated as "the official cultural practice of all of our institutions. It is the only commonly understood discursive and representational system" (Rapping, 1992, p. xxxiii). According to Rapping, the radical potential of TV movies is often capped by their participation in realist portrayals that do not challenge the stereotypes of bourgeois society. This is aggravated by the common assumption that realist texts do not represent reality in a mediated manner. In such a system, the political conditions of life in patriarchal, heteronormative societies are assumed as given, and the status quo is upheld as a natural, not socially constructed, condition. It is clear how an attachment to the structures of realism effectively handicaps any discussion of cultural difference, which would seek to undermine the presupposed "naturalness" of imposed systems of distinction such as patriarchy, heterosexuality, and white supremacy.

Ultimately, then, TV movies present a highly contested space for the introduction of queer elements. As a women's genre segregated from the rigors of high culture, they offer greater flexibility for reading "against the grain" because the critical space surrounding them has not been exhausted by decades of academic contestation. Yet, as participatory in the normative patterns of televisual discourse, the idealism of American family life, and the roles implied for women by these participations, TV movies also resist their potential as sites for the production of radical queer meaning. In the last decade, however, TV movies appear to have increasingly shifted from their identification with a traditional picture of idealist, realist American life–largely through the incorporation of queer elements. Forced open by queer activism and protest, this small representational space has allowed lesbian content a passageway into mainstream American culture. Indeed, TV movies have become a definitive venue for lesbian representation in the mainstream media. Although Rapping points out that TV movies depicting queer experience have been produced since the mid 1980s (Rapping, 1992, p. 102), the structure of recent television movies has expanded upon these earlier models and initiated new forms of sexual storytelling carrying different messages about what queerness might mean in American society. An appropriate examination of these "new" TV movies requires a discussion of the problematics of lesbian representation in contemporary popular media.

Monique Wittig (1993) provides valuable theoretical insight into the difficulties of constructing a discourse around lesbianism and the television movie as it has been generically developed. Wittig establishes an argument that uncovers the political and economic fabrication of the roles "man" and "woman," claiming that an idyllically lesbian society would destroy the "artificial (social) fact constituting women as a 'natural group'" (Wittig, 1993, p. 106). Wittig is (in)famous in her assertion that lesbians are, in fact, not women, "for what makes a woman is a specific social relation to a man, a relation that we have previously called servitude . . . a relation that lesbians escape by refusing to become or to stay heterosexual" (Wittig, 1993, p. 108). Though it is doubtful that a lesbian identity provides a complete escape from the various institutional servitudes of patriarchy, Wittig's claim that lesbians are *not* women has been upheld by media depicting lesbians as something other than women and therefore impossible to assimilate into mainstream culture. Yet the issue of lesbianism is still embedded within the TV movie's larger concern with the "concept of woman" as a social problem. In order to include lesbians in their subject matter as anything more than a threat to social integration, TV movies must somehow transform lesbians into women, a project congruent with their preoccupations over sexual roles and the defense of the nuclear family.

Lesbian representation in popular media is also specifically crippled by the boundaries of realism, the preeminent symbolic system of televisual discourse. Jill Dolan (1989) posits that realism perhaps *cannot* accommodate feminism, people of color, or lesbians and gay men, since it operates as a "mimetic site" for the arrangements of dominant ideology. Dolan writes that, "because the signs of sexuality are inherently performative, the assumption of heterosexuality prevails unless homosexual or lesbian practice is made textual" (Dolan, 1989, p. 149). Given the heteronormative qualities of realism, Dolan acknowledges that the introduction of an actual lesbian identity in such texts must necessarily fragment the organization of realist representation. "The lesbian performer breaks the identification pacts of realism, in effect, to lead people in a discussion of what they *see* in representation" (Dolan, 1989, p. 152). In order to depict a "lesbian verisimilitude" in keeping with the growing public awareness of lesbian realities, the successful "queer" TV movie must therefore deviate at least slightly from the tradition of bourgeois realism.

The concessions queer identity has been forced to make in order to enter the discourses of popular culture have permitted an external understanding of "queerness" that is recognizable to the heterosexual viewing audiences of television and TV movies. It is this "queerness," reinforced by the heterosexist doctrine of psychoanalysis, which Valerie Traub (1995) refers to when she writes that, "despite the historical variability in what a 'lesbian' 'is,' a particular construction of the 'lesbian' has achieved the reified, if unconscious, status of the 'real'" (Traub, 1995, p. 115). This construct of the "real" and perceivable (i.e., heterosexualized or "fake") lesbian is achievable only in a binary system of gender that upholds the psychoanalytic concept of sexual difference and therefore conflates gender identification and erotic desire. Traub writes that even the most radical feminist theorizing couched in this psychoanalytic model will result in "an elision of erotic practice"

(Traub, 1995, p. 117). "Lesbians" who have achieved the status of the "real" in entering realist and heterosexist texts are then forbidden from expressing queer sexuality, and are left with only two ideological positions to occupy: (1) wanting to be a man, or (2) wanting to be the *other* (desired) woman. These two heterosexualized posts are actually inversions of one another: based on the assumption of the "lesbian's" gender dysfunction, the oppositional structure of the binary "preserves the seemingly essential heterosexuality of desire" (Traub, 1995, p. 118). The gender dysphoria of the faux "lesbian" in realist representation functions as only a perversion of heterosexual norms, and therefore "subverts, but cannot *overturn*, the hegemony of binary codes" (Traub, 1995, p. 120). This subversion still is dependent upon the bipolar terms of gender asymmetry for its visibility and, therefore, its very existence within the text. Traub makes two points essential to understanding the bargaining process at play between the entrenched conventions of genre and the new push for queer authenticity in telefeatures that attempt to represent the lesbian as more than "real"–as actual:

> What is at stake, then, in the representation of "lesbian" desire in [heterosexual texts] is the very intelligibility of the narrative; correspondingly, what is at stake in [those same texts] is the very possibility of "lesbian" representation within a masculinist and heterosexual field. (Traub, 1995, p. 123)

The conflicts at work in this troubled relationship between the representational traditions of television and any form of queer agency have resulted in a number of televisual tropes that serve to solve the problems of queerness as it enters the popular media. Rose Collis (1994) writes that television will always work to keep too-queer lesbians "out of the mainstream" (Collis, 1994, p. 120), and goes on to highlight the various coping strategies television has developed in order to admit yet simultaneously reject lesbian representations. Lesbian and other "queer" appearances are ghettoized into low culture programs such as talk shows, soap operas, and TV movies, which are separated from the centers of media control and therefore allow what has been made uncomfortably visible to remain largely invisible to the dominant culture. Representations of lesbians on television are almost always presented as "issues" or "subjects up for analysis," and can be classified into two primary categories: the "coming out" story and the "lesbians-as-criminals" story. Both of these stories, which make up our basic cultural myths about lesbians, depict queer women as they are perceived from within the master discourse of heterosexuality. Collis writes that "what television has accurately represented (though not always with total sympathy) are the problems lesbians face when other people . . . perceive their sexuality as a problem" (Collis,1994, p. 129). Collis asks television viewers to think twice about what actually has been accepted as "positive" lesbian representation, since merely escaping (or being banished) from the televisual discourse does not constitute a happy ending for characters who–in a vicious circle–are forever forced to seek inclusion in the mainstream culture. Lesbian television characters are disposable in their personification of social issues that must be solved, and are easily written out of shows because their queer following is generally unimportant to the industry's corporate sponsorship. Collis is blunt in her assertion that lesbians on TV "don't have proper sex," and emphatically reminds us that "TV dykes don't have happy endings" (Collis, 1994, p. 139, p. 128). The article is summed up by stating that "the harsh fact has to be faced: mainstream television, especially in America, will never *properly* (read "authentically") put lesbians in the *picture* (read "realist heterosexual text")" (Collis, 1994, p. 139, emphasis and parentheticals added). However, despite Collis's fatalism, the recent explosion of TV movies focusing on lesbian representation provides a space in which to further engage the conflicts of queer representation in the mainstream media. These TV movies have been pushed by the inclusion of "queerness" to redefine the very structure of the genre. As "queer" images begin to enter the greater body of televisual discourse, we may expect television to eventually undergo similar shifts in the overall logic of its representation.

In 1994, the Lifetime Network produced a TV movie that altered the rules governing the representation of lesbians in popular media by allowing a central queer subjectivity capable of expressing desire into the televisual discourse. *Serving in Silence: The Margarethe Cammermeyer Story* employed well-known Hollywood actress Glenn Close to play Cammermeyer, the highest-ranked official to ever be removed from military service for being homosexual. Far from being a radical depiction, the telefeature presented Cammermeyer's story as recuperative of the nation's integrity–she is coded as a hero because, despite her difference, she only seeks to serve her country. The figure of Cammermeyer is presented as tough yet nurturing; her roles as a mother who holds together a broken family and a Colonel who commands an Army unit are effectively collapsed into one another, providing a story suspiciously free of gender dysfunction. Although Cammermeyer is a divorced woman who plays tackle football with her three sons and acts as the family's breadwinner, she still asserts the importance of hetero-sexual family structure by stating that "boys need a father" as well. The central conflict of the movie is presented by Cammermeyer's love inter-est, a character named Diane, who asks Margarethe if she "wants to have it all." As the movie will indicate, it will be impossible for Cammermeyer to simultaneously retain her family, her military career, and her growing sense of queer identity. In order to remain within the category of the representable "woman," the figure of Cammermeyer must concede a radical queer identity and a career as an Army General to the more conventional roles of mother and medical nurse. It is only in this context that Cammermeyer appears feminine enough to occupy the positions of woman and mother required by the TV movie genre.

Cammermeyer's queerness is normalized by the movie's processes of enclosure, sanitation, and heterosexualization. The lesbian aspects of the plot are developed in the private space of the home and are largely isolated from the outer political world. Cammermeyer may be one of the first lead characters to assert her lesbianism without disgust or re-morse, but the expression of her queerness is hampered by a sanitized eroticism. Margarethe and Diane are allowed only cursory hugs and one chaste kiss, which occurs late in the movie. The telefeature also permits male, heterosexual authority to remain in an unchallenged position of power. Cammermeyer must "out" herself to male figureheads in order to ensure the safety of her job, her family, and her roles as mother and daughter, and her legal suit against the Army is presented as a struggle to serve, not overthrow or resist, the structures of governmental control. Although she says, "I'm not a hero," the movie informs us otherwise, for in ultimately choosing to rebuild the family ("queer" as it may be) and relinquish her military career, Margarethe upholds the more traditional values of social integration and controlled sexuality.

Ultimately, *Serving in Silence* provides an unprecedented space for queer representation in televisual discourse by coding queerness as compatible with family life in America and with service to the na-tion. Despite its attempt to solve queerness by presenting it as heroic martyrdom, the movie must also make concessions to queer identity by presenting a de-heterosexualized, if broken, family structure and by permitting the entry of lesbian desire into representation. However, dif-ference is no longer the subject of debate: the political struggle waged by Cammermeyer as a lesbian fighting her removal from service is translated into the conduct of a dutiful soldier who privileges honesty and directness over personal pleasure. The final scenes of *Serving in Si-lence* present Cammermeyer delivering a speech at a gay pride rally, claiming that queer people deserve equal rights because they are "just the same" as heterosexuals. This speech implies that queer people do not represent a threat to heteronormative rule, but wish to serve the same systems that oppress them. "All we ask, is that you let us," Cammermeyer pleads.

In the year 2000, another groundbreaking television movie about lesbians was released to unprecedented success. Home Box Office, free from primetime censorship and with access to Hollywood "star power," aired *If These Walls Could Talk Two*, a three-part film written in the traditional vein of TV movies concerned with resolving social issues. The "issue" in *Walls* is, in fact, lesbians, but the principal concern of the movie is the crisis of the American family. *Walls* is a slick vehicle that presents an insular and lesbianized universe, unfolded through three generations of lesbian women living in the same house (*Walls* departs quite perceptibly from the tenets of realism, as any viewer will recognize). In its progression forward through a retroactively imagined history, *Walls* endeavors to prove the salience of lesbian political "progress" by depicting lesbian reality as more and more fully absorbed into the structure of the American family.

Walls' first section opens with documentary footage arranged in a montage showing actual lesbians engaged in protest and activism. Although this footage supports the movie's claim to a visible lesbian history, the images also serve to somewhat undermine the film's romanticized portrayal of queer integration. The section's story features the elderly characters Edith and Abby as a closeted lesbian couple living in the early 1960s. The two are a perfect–and deconstructivist–picture of domesticity, but when Abby falls from a ladder and then dies at the hospital, their home is invaded and "vanished" by the heterosexual discourse of the nuclear family. When Abby's relatives come to arrange the funeral and estate proceedings, Edith is forced to hide the sexual nature of her and Abby's relationship. Powerless to fight the authority of the documents and files that deny her desire and her very existence, Edith is forced to surrender her home, literally disappearing from the *mise-en-scène*. It is this same authority that must grant the existence of queer women before lesbians will be permitted to participate in the structure of the American family. The light patches on the house's yellowed walls, left where pictures of Edith and Abby used to hang, are the only indication of a possible future where queers might enjoy freedom–although the actual parameters of this "freedom" grow more dubious as the film progresses. The remainder of the movie will be concerned almost entirely with the development of a future lesbian "family space" that permits access to the dominant discourses of capitalism and heterosexual privilege, and acquires its legitimacy from its increasing acquiescence to and resultant visibility within the systems of hegemony.

The second section of *Walls* performs a massive leap in representation: the 1970s have arrived and feminism has descended upon suburbia. Through the influence and political clout of the feminist movement, a new form of "lesbian home" has become possible. Linda lives with a group of young and rebellious queer women who owe their identities largely to the protest dynamics of second-wave feminism. Although this group incorporates more than one type of difference–a black dyke is included–racial and class conflicts are collapsed into queerness in order to maintain the integrity of the movie's binary oppositions. When the group of girls is kicked out of the feminist collective on their college campus, it becomes evident that feminism and lesbianism are not as easily reconciled as might be assumed. Linda, bored with the ennui of the gay ghetto and sickened by her friends' small-minded denigration of working-class butch/femme roles, stays behind at a lesbian dive bar and meets a handsome butch girl, Amy, who wears a tie and rides a motorcycle. Linda and Amy pursue and ultimately succeed in having a romantic relationship, despite the negative reactions of Amy's friends and the couple's conflicting identities and ideas of what it means to be queer.

This section of *Walls* is the most progressive in its exploration of the differences dividing lesbian experience, and is in effect "paid for" by the greater conservatism of the other sections. The lesbian queerness we are so familiar with today has become a recognizable form of identity, but constantly threatens to overflow its definitions as middle-class, upwardly mobile, white, and assimilationist. The movie defuses the fragmentary threat of the differences pooled within queerness by turning Amy's butch identity–a political statement undermining the biological essentialism of heterosexual roles–into a simplified wish to be "comfortable." "This is who I am," Amy says. Amy's "butchness" must be contained because it serves as tangible evidence of the radical potential of queerness to unravel the assumptions of a patriarchal, heterosexist society. Even though the educated, feminist lesbians are grudgingly admitted into the "square" environment of the suburbs, in Amy's house on working-class 4th street we are again reminded of the basic incongruencies of queer and heterosexual space. Awaking from a night spent in Amy's bed, Linda opens the bathroom's medicine cabinet to find a tiny, faded picture of Amy's family. The photo is a testament to the unassimilatable quality of Amy's identity, and an injunction to queerness, which must abandon its more radical capacities in order to be welcomed into the heterosexual discourse of home and family.

The third section of *Walls* resolves the tensions surrounding queerness by placing a lesbian couple, through a number of ideological compromises, at the center of the American nuclear family. An upper-class pair of white lesbian women (Ellen DeGeneres and Sharon Stone) are "married" and pursuing a pregnancy. The segment consists largely of the women mapping ovulation cycles, choosing, buying, and administering sperm, and watching children play on a playground while holding hands. This section is particularly interesting–but ultimately disappointing–in its representation of lesbian conformity to the patterns of heterosexuality as "progress" in the name of equal rights. These lesbians have become acceptable to the hegemonic discourse through a mimetic exercise of heterosexual privilege: they are married, wealthy, procreative, and invested in the ultimate importance of the nuclear family. The political agency of queer sexuality has been relinquished for a greater participation in the assimilative powers of capitalism: the lesbians must purchase a product (sperm) that will compensate for their familial inadequacy and allow them to achieve full adaptation to the nuclear family model. In an extremely dissatisfactory scene, Stone asks DeGeneres whether she thinks it is right for them to have a child because it will be punished for not having heterosexual parents. DeGeneres responds with the weak answer that world has always "gotten better," and that all they have to do to fight oppression is "love their kid." The political movement that fueled Linda's surrogate family in section two has been completely subsumed by the reified position of the heterosexual family and the sacrosanct status of motherhood. The family is no longer required to be physically heterosexual, but what does this matter if the political result is the same? Romantic love, not the erotic resistance Audre Lorde wrote about (Lorde 57), has become the answer to queer oppression.

As an unprecedentedly popular text about lesbian "issues," *Walls* encouraged the production of more TV movies depicting lesbian experience by proving the subject matter both safe and profitable. It is not surprising, then, that in 2001 Lifetime Television produced another movie depicting a central lesbian character. *The Truth About Jane* is a coming-out story, one that explores the difficulties of being a white, middle-class, homosexual teenager at the same time that it attempts to recuperate and vindicate the selflessness of heterosexual motherhood. Principally, the movie questions how far "family values" and the definition of motherhood can be made to stretch in the face of difference. The movie's central character, Jane, narrates the story of her own coming-out, which is set against the backdrop of an idyllic suburban community. Fifteen and just entering high school, Jane realizes she is "gay" when she develops a crush on another girl, Taylor, who is openly queer and comes from a broken family living on the "wrong side" of town. Seeking advice, Jane confides in her mother's queer and black friend, Jimmy, who accepts and comforts her. However, when Jane and Taylor are caught kissing, Jane's queerness becomes an issue in the larger community of her high school and in her family. Jane comes out to her parents, but their reaction is negative, and her mother is revealed to be painfully homophobic. Rejected by her peers, her family, and Taylor–who breaks up with her over the scandal–Jane seeks support from her English teacher and guidance counselor, Ms. Wolcott, who is also a lesbian. After her parents inform her that they have decided to send her to boarding school, Jane runs away to Ms. Wolcott's home, where she admits that she has been contemplating suicide. The movie then becomes a conflict between Jane's lesbian identity, which cannot be erased, and her mother's expected roles as a nurturer of children and the center of the family. Ultimately, Jane's mother must choose to accept Jane's homosexuality in order to preserve the family structure and ensure the selfless qualities traditional to motherhood.

As a TV movie, *The Truth About Jane* is perhaps the most progressive bargain that has been struck between the conditions for a queer subjectivity and the limitations of the TV movie genre. The film grants its main character sexual desire, erotic expression, agency, and a powerful voice that controls the terms of the televisual discourse. Shown at the cusp of her "womanhood" as an adolescent, Jane is allowed to engage in behaviors (sleeping with Taylor, dancing with other women at a gay bar) that code her as something other than a properly defined woman. The movie's depiction of difference is also more sophisticated than anything portrayed in *Serving in Silence* or *Walls*. The character of Jimmy, whose difference is compounded by both racialized and homosexual identities, is even "more different" than Jane, and signals her eventual assimilation into the suburban community. Responding to Jane's insistence that everyone "hates" her, Jimmy quips, "Honey, I'm black *and* I'm gay. They hate me way more than they hate you." The movie's repressive suburban setting, which functions as an extension of the heterosexual family structure, is presented as having spaces that can still be occupied by queer people–a gay bar, a PFLAG meeting, and a pride rally.

The progressive aspects of *The Truth About Jane* are matched, however, by a decidedly conservative strain that forces the audience to accept a situation in which queerness can never really exist outside of the family structure. The figure of the TV movie lesbian, already situated in relation to the heterosexualized roles of wife, mother, and daughter, becomes a "tragedy" that provides an opportunity for the fragmented family to reconstruct itself through the traditional values of love and social integration. Although this family is no longer thoroughly heterosexual, it still conforms to the nuclear patterns that support capitalist social structures. It is clear that Jane will *not* be okay until her family accepts the "truth" about her, but it is doubtful how complete that acceptance can become. This reluctance, or inability, to fully grasp a queer subjectivity, is consolidated in the character of Jane's mother. By the end of the movie, she has begun to act supportive of Jane, but admits to her husband that she is still "just pretending." The deeply homophobic qualities of this character carry very negative implications for American motherhood, a role that the movie is compelled to uphold in its generic address to a female consumer audience. *The Truth About Jane* is, like *Serving in Silence* and *Walls*, devoted to saving an American family under the threat of difference, but it achieves this through a reification of motherhood as essentially heterosexual and fettered by its centralized, domestic position. The movie reinscribes the absolute and static value placed on selfless motherhood in heterosexual society: Jane's mother finally accepts Jane's sexuality not because it is the socially just thing to do, but because Jane is her child and therefore is owed unconditional love.

The restraints of the media in general, and the TV movie in particular, guarantee a picture of queerness that enters the televisual discourse only as much as it concedes the radical and deconstructive potentials of queer identity and desire, which in their "authentic" form threaten to entirely rupture the mimetic function of the media's images. Given the critical success of past TV movies portraying lesbian characters, it is no surprise that Lifetime now releases at least one queer "issues" movie per season. It is also no surprise that two of these recent TV movies, *What Makes a Family* (Brooke Shields plays a lesbian mother fighting to adopt her lover's child) and *An Unexpected Love* (a divorced woman falls in love with her female boss and must reconcile her family), follow the same patterns of representation–they collapse the political struggle for queer "rights" into a highly personal fight to participate in the American family and its procreative privileges. The result is a simplified and largely apolitical commentary on the universalizing power of romantic love and the right of all people to "be who they are." Such narratives are undoubtedly "progressive" in their attempts to incorporate queer identity into mainstream media, but such incorporation also requires a process of containment and erotic sanitation. This process guarantees media representation to only specific deployments of queerness that do not challenge heteronormative arrangements of power and do not actively threaten the political status quo.

The erotic sanitation imposed upon queerness by televisual discourse is far more complicated than the negative injunction to merely "disappear." The process is instead a hegemonic negotiation in which the internal contradictions of heteronormative systems of power are translated into queerness itself. Thus, television movies about "queerness" become movies about something else entirely, whether this is the breakdown of the nuclear family, American motherhood, or the difficulties of heterosexual marriage. These texts *generate* queerness as a problem through which they imagine that community and heterosexual integration might be restored. This utopian project is exemplified by HBO's recent television movie *Normal*, which seeks to explore transsexual issues in the same manner that *Walls* explored lesbian identity. As an exercise in the powers of progressive assimilation, however, *Normal* fails utterly–chiefly because its narrative relies upon the formula established by the previously discussed TV movies. Although attempting to depict a man's struggle to retain his marriage and children while accepting and realizing his new female identity, *Normal* ultimately becomes a story about the survival of marriage and family despite the death of maleness. As the main character's transition progresses, s/he is pushed further into the domestic home and away from the public sphere, prohibiting the movie's own inclusive pretext. Although the movie's family manages to remain intact, the televisual community vanishes. The implied result is that transsexuality is incommensurate to the movie's integrative intentionality.

The failure of *Normal's* effort to thematize transsexuality as a model for the restoration of normative community structures is due in large part to the inability of the televisual text to depict queerness as a shifting *plurality* of sexual and gender identities. Relying upon the formula established to depict an assimilated lesbianism prevents the movie from presenting transsexuality as *other than* lesbianism. It is precisely because the film's main character *rejects* a lesbian identity that the movie cannot go forward in its assimilative logic. Indeed, *Normal* ends abruptly on the night before the main character's "operation" is to take place, shutting out all the uncertainties of what might follow (including non-heterosexual sex). By foreclosing these possibilities, the movie commits itself to a conservative framework in which "transitioning" is just another word for the end of manhood. The assumption that *Normal's* inability to signify transsexuality is a sign of that identity's "truly radical" nature would be a dubious one: instead, the movie seems to suggest that there is only a certain amount of "room" for queers in the mainstream media, and that some forms of queerness (i.e., white lesbianism) are being used as tools to effectively erase others.

If we are to view televisual texts as indicators of how social dynamics are ordered, as examples of how we wish American culture to be elaborated and what stories we wish to be told, then the outcome of this process of queer media containment is necessarily bad for the transgendered and transsexual, for gender nonconformists, for working class and homeless queers, rural queers, queers of color, disabled queers, fat queers, nonmonogamous queers, and for queers who do not wish to form "marriages" or nuclear-style families. Although the heterosexist modes of media representation are momentarily compromised by the presence of queerness, they are ultimately recuperated by a chain of significations that allows the media to entirely skirt the larger question of sexual freedom in American society. The "queer" becomes recognizable as a media object and a capitalist subject, rather than as a sexual and political agent. If popular media, and TV movies in particular, now serve as our chief method of social self-reflection–of picturing the parameters of our own "imagined community" (Anderson, 1991)–then this is a community still resistant to queer enfranchisement and the profusion of sexual and political possibilities a public and active queerness might prompt in American culture. These media products may make queerness a little more comfortable, a little more "tolerable" for the larger population, but it is doubtful that they somehow make American politics any more democratic or socially just. To rely on the bounded framework they allow for queerness would be to preempt the movement for our own liberty.

ISSUES AND QUESTIONS FOR FURTHER CONSIDERATION

1. If queer textuality is often conservatively mediated by processes of representation under neoliberal democracy, then is a critique of democracy itself a necessary step in the development of a queer theory that would support substantial sexual freedoms?

2. If queer people cannot entirely remove themselves from neo-liberal hegemony through subcultural affiliation, then will the involuntary mainstreaming of queerness always take place through narratives of containment? How is queer agency to be retained when queerness has become a commodity?
3. How are queer people to critically manage the reality that we often find pleasure in media representations of queer life that ultimately recommend our containment and sterilization?

REFERENCES

Anderson, Jane. (Director). (2003). *Normal*. [Film]. (HBO Home Video).

Anderson, Jane, Coolidge, Martha, & Heche, Anne. (Directors). (2000). *If These Walls Could Talk 2*. [Film]. (HBO Home Video).

Anderson, Benedict. (1991). *Imagined Communities: Reflections on the Origin and Spread of Nationalism*. New York: Verso.

Bleckner, Jeff. (Director), & Heus, Richard (Producer). (1994). *Serving in Silence: The Margarethe Cammermeyer Story*. [Film]. (Tri Star Television, Inc. 10202 West Washington Blvd. Culver City, CA 90232-3195).

Collis, Rose. (1994). "Screened Out: Lesbians and Television." *Daring to Dissent: Lesbian Culture from Margin to Mainstream*. Ed. Liz Gibbs. London: Cassel, 120-146.

Dolan, Jill. (March, 1989) Breaking the Code: Musings on Lesbian Sexuality and the Performer. *Modern Drama*. Vol. XXXII, No. 1. 147-158.

Lee, Rose (Director), & Adelson, Orly (Producer). (2001). *The Truth About Jane*. [Film]. (Hearst Entertainment).

Lorde, Audre. (1984). Uses of the Erotic: The Erotic as Power. *Sister Outsider*. Freedom: The Crossing Press, 53-59.

Rapping, Elayne. (1992). *The Movie of the Week: Private Stories, Public Events*. Minneapolis: University of Minnesota Press.

Traub, Valerie. (1995). The Ambiguities of 'Lesbian' Viewing Pleasure: The (Dis)articulations of *Black Widow*. *Out in Culture: Gay, Lesbian, and Queer Essays on Popular Culture*. (Eds.) Corey K. Creekmur and Alexander Doty. Durham: Duke University Press, 115-136.

Wittig, Monique. (1993). One is Not Born a Woman. *The Lesbian and Gay Studies Reader*. (Eds.) Henry Abelove, Michele Aina Barale, and David M. Halperin. New York: Routledge, 103-109.

"My Spirit in My Heart":
Identity Experiences and Challenges Among American Indian Two-Spirit Women

Karina L. Walters
Teresa Evans-Campbell
Jane M. Simoni
Theresa Ronquillo
Rupaleem Bhuyan

Karina L. Walters, PhD, holds the William P. and Ruth Gerberding University Professorship at the University of Washington (UW) School of Social Work (SSW). She directs the Native Wellness Center there and conducts research on cultural strengths in indigenous populations that buffer the effect of historical trauma and discrimination on health.

Teresa Evans-Campbell, PhD, is Assistant Professor at the UWSSW where she co-directs the Institute for Indigenous Health and Child Welfare Research and researches the effects of historical trauma, including boarding school experience, on indigenous families.

Jane M. Simoni, PhD, is Associate Professor in the UW Department of Psychology. She teaches a course on minority mental health and conducts research on medication adherence and psychosocial challenges among individuals living with HIV/AIDS.

Theresa Ronquillo, MSW, is a doctoral student at the UWSSW where she works as a research assistant with the HONOR Project, a study of two-spirit health. She is interested in processes of identity and decolonization among indigenous and other minority communities.

Rupaleem Bhuyan, MA, is a doctoral candidate at the UWSSW. Her research interests center on developing culturally competent theories and practices through participatory action research with minority communities.

Address correspondence to: Karina L. Walters, University of Washington School of Social Work, 4101 15th Avenue NE, Seattle, WA 98105 (E-mail: kw5@u.washington.edu).

[Haworth co-indexing entry note]: "'My Spirit in My Heart': Identity Experiences and Challenges Among American Indian Two-Spirit Women." Walters, Karina L. et al. Co-published simultaneously in *Journal of Lesbian Studies* (Harrington Park Press, an imprint of The Haworth Press, Inc.) Vol. 10, No. 1/2, 2006, pp. 125-149; and: *Challenging Lesbian Norms: Intersex, Transgender, Intersectional, and Queer Perspectives* (ed: Angela Pattatucci Aragón) Harrington Park Press, an imprint of The Haworth Press, Inc., 2006, pp. 125-149. Single or multiple copies of this article are available for a fee from The Haworth Document Delivery Service [1-800-HAWORTH, 9:00 a.m. - 5:00 p.m. (EST). E-mail address: docdelivery@haworthpress.com].

Available online at http://www.haworthpress.com/web/JLS
doi:10.1300/J155v10n01_07

SUMMARY. Many Native women embrace the term *two-spirit* to capture their sexuality and gender expression. By analyzing the narratives of five two-spirit women who are Native activists, we explored contemporary understandings of the concept and what it means for Native communities. The incorporation of the identity within indigenous worldviews, its manifestation in terms of (be)coming out, and the triple stressors of heterosexism, racism, and sexism emerged as key themes. *[Article copies available for a fee from The Haworth Document Delivery Service: 1-800-HAWORTH. E-mail address: <docdelivery@haworthpress.com> Website: <http://www.HaworthPress.com> © 2006 by The Haworth Press, Inc. All rights reserved.]*

KEYWORDS. American Indian/Alaskan Native, lesbian spirituality, lesbian women of color, two-spirit, sexual orientation and discrimination, identity development, qualitative research

I feel as though being a queer Indian is the hardest job in the world . . . you have a colonized situation and dissolution of traditional ways–it's hard to be queer and Indian. (Maxine, a two-spirit Native activist)

Historically, Native societies incorporated gender roles beyond *male* and *female* (Brown, 1997; Lang, 1998; Little Crow, Wright, & Brown, 1997). Individuals embracing these genders may have dressed; assumed social, spiritual and cultural roles; or engaged in sexual and other behaviors not typically associated with members of their biological sex. From the community's perspective, the fulfillment of social or ceremonial roles and responsibilities was a more important defining feature of gender than sexual behavior or identity. Although there were exceptions, many of the individuals who embodied *alternative* gender roles or sexual identities were integrated within their community, often occupying highly respected social and ceremonial roles.

Western colonization and Christianization of Native cultures, however, attacked traditional Native conceptions of gender and sexual identity. The colonizing process succeeded in undermining traditional ceremonial and social roles for two-spirits within many tribal communities, replacing traditional acceptance and inclusivity with shaming condemnation (Tinker, 1993).

From within the academy, anthropologists have sought–unsuccess-fully–to understand the historical status of indigenous peoples who lived with more fluid gender and sexual expressions (Farrer, 1997). As Blackwood (1997, p. 285) explained, "The critical importance of biol-ogy to Western constructs of gender meant that White scholars were rarely able to separate biology from gender successfully." The label of *third gender* they proposed is based on the Western binary system of gender and diminishes the complexity of multi-gendered statuses and expressions. The term *berdache* is offensive because of its colonial ori-gins and purely sexual connotations: it is a non-Native word of Arabic origin (i.e., *berdaj*), which refers to male slaves who served as anally re-ceptive prostitutes (Jacobs, Thomas, and Lang, 1997; Thomas & Jacobs, 1999). More contemporary anthropologists created the terms *women-men* and *men-women*, which are similarly deficient (Lang, 1998).

Native activists emerged with a term of their own–*two-spirit*. Adopted in 1990 at the third annual spiritual gathering of lesbian, gay, bisexual, and transgender (LGBT) Natives, the expression derives from the Northern Algonquin word *niizh manitoag*, meaning *two spirits*, and refers to the inclusion of both feminine and masculine components within one individual (Anguksuar, 1997). The term *two-spirit* is used currently to reconnect with tribal traditions related to sexuality and gen-der identity; to transcend the Eurocentric binary categorizations of ho-mosexual vs. heterosexual or male vs. female; to signal the fluidity and non-linearity of identity processes; and, to counteract heterosexism in Native communities and racism in LGBT communities (Walters, 1997; Walters et al., 2001).

Blackwood (1997) suggested moving beyond labeling and classify-ing two-spirits to considering two-spirits as part of lived, contemporary human culture, situated within social relations that are negotiated and contested by family, community, and historical interpretations. She fur-ther recommended that extending the analysis of two-spirit gender into the realm of social relations and asking how two-spirit people position themselves in relation to other Natives as well as to White LGBT groups and individuals.

Toward these ends, we present in this paper experiences, perceptions, and challenges regarding the adoption of a two-spirit identity among Native women based on data from a large-scale national study of two-spirit health (i.e., the HONOR Project). Working in concert with local and regional two-spirit communities and Native agencies, HONOR Project staff conducted over 60 in-depth interviews with two-spirit leaders and activists covering topics from identity to community strengths to health concerns. Consistent with narrative and indigenist research methods, the qualitative interviews provided opportunity for two-spirit leaders to give their *testimonios*, a type of oral history and life story as two-spirit leaders and women (Bishop, 2005; McMahon & Rogers, 1994; Tuhiwai Smith, 2005). Interviewers did not focus on eliciting factual historical data; rather, they aimed to uncover the meanings that familial, spiritual, communal, and historical events have in shaping identity and quality of life for two-spirit women. The five two-spirit women whose narratives we review here range in age from late 20s through late 50s and represent considerable tribal diversity. To protect confidentiality, pseudonyms are used and limited tribal and other socio-demographic information is provided. Many of the quotes presented here were edited for readability and grammatical correctness.

"IT'S ON A MORE DEEPER SPIRITUAL LEVEL": DE-CENTERING SEXUALITY AND CENTERING SPIRITUALITY IN TWO-SPIRIT IDENTITY

For Native persons, indigenous worldviews, including the centrality of spirituality, and ways of relating form the core of any behavioral expression. Indigenous traditional worldviews recognize the interdependency among humans and nature, the physical and spiritual worlds, the ancestors and future generations–connections that bind all living beings in spiritual ways. It is not surprising then that the term *two-spirit* is connected to traditional spiritual values and extends beyond the mainstream focus of sexual orientation as rooted in sexuality. Alex Wilson (1996), a two-spirit woman activist and educator, wrote that the term two-spirit "proclaims a sexuality deeply rooted in our own cultures. Two-spirit identity affirms the interrelatedness of all aspects of identity, including sexuality, gender, culture, community, and spirituality" (p. 303). The women in our study concurred with this perspective.

Being two-spirited, kind of goes beyond my sexuality. I am attracted to women, prefer to be with a woman, but it also is more about who I am as a person . . . there's a spiritual side to it . . . there is a spiritual side that I just can't find words for. (Sandy)

For me, I look at the word and I hear the word two-spirit, I look at the spiritual component of that, and I have to really say if I use this, what does this mean to me on a spiritual level, what is my identity to this on a spiritual level? . . . I say I'm a spiritual woman walking a spiritual way of life. You know, that's how I really want to be seen, that's how I really want to be known. Not as necessarily the two-spirit woman, but a woman who's walking the spiritual path and struggling on that spiritual path but learning to walk it and to embrace it and be a spiritual kind of woman, but it's not gender specified . . . so anyways, so yeah. (Roberta)

"BEING RESPONSIBLE TO THE PEOPLE": TWO-SPIRIT IDENTITY MEANS SERVING THE COMMUNITY

Indigenism values familial, communal, and ancestral roles and responsibilities. Indeed, all of the two-spirit women talked about the general importance of their roles as community caregivers.

I just feel like I have this responsibility to community that I have to fulfill and that's just a part of me and I feel like I'm–that's just something I naturally gravitate to, even if I consciously don't feel like I want to do that, you know, it's like I've gone through periods where I needed to take care of myself and I need to do this and work at this job or go to school or whatever, but it's like this community always seems like it takes priority to me, that I need to be something no matter how small it is . . . (Winona)

Additionally, many connected their fulfillment of these roles specifically to their identity as two-spirit women.

I believe that there have always been roles and I believe that each nation has had names for people like us. And it was just a way of life. There was no–my belief is that we didn't have to explain who we were, we didn't have to justify our existence, we were just who we were, part of our community and part of our village and those roles that we took on were just a natural aspect of who we were . . . And within myself today I see it happening. I don't have to think about it, I just fall into some role. I don't say–yeah, I'll do that! You know? It just happens to be. I might be there at the right time, the right place. I always believe that there have been roles for us. And those roles, either spoken or unspoken, [involve] just naturally filling in those gaps. (Roberta)

The traditional sacred ceremonial roles for two-spirit women have expanded in contemporary space to include political organizing and engaging in legal battles for indigenous sovereignty. As Maxine indicates, these roles often are perceived as a central organizing component of two-spirit identity.

> I feel like the gift that I can try and give back to the Native community is um speaking out and saying things I've said and the work that I've done for treaty rights. And, of course, I'm one of many, many people. I mean it's not like I'm center of anything. I'm just doing what I see to do . . . that's where my heart is, you know, in trying to do that kind of work and make people's lives easier. Like I said, sending stuff to the rez and you know. . . . I don't have any traditional knowledge, I can't speak my language fluently, I'm not a very useful Indian [laughs] for Indians except for what I can do you know to help people who have less than myself so that's kind of what my center is.

Overall, this approach to conceptualizing a two-spirit identity is similar to how the women view their Native identity: multifaceted, involving spirituality, and manifesting in socially sanctioned behavior.

> My great-grandfather said to me once that being Indian isn't being on a piece of paper; it's spirit. So I don't think for me that's what it's always been about. My identity isn't on a piece of paper, isn't on a tribal card, isn't a number–it's me and my spirit in my heart and the way I choose to live my life. So that keeps me grounded in what I do. (Roberta)

"WE NEEDED TO KIND OF DEVELOP OUR OWN IDENTITY": COUNTERING OPPRESSIVE DOMINANT DISCOURSES WITH A TWO-SPIRIT IDENTITY

Many of the women spoke of how the term two-spirit emphasizes the importance of indigenous worldviews, histories, and experiences in the face of White hegemony in the mainstream LBGT community. Winona commented on how the two-spirit term has come to represent a form of indigenous resistance:

I usually use the term two-spirit . . . most people felt a lot of alienation from the White gay/lesbian community and really–I don't want to make it seem like it was reactionary to that–it really felt like we needed to kind of develop our own identity outside of that prejudice and I think the term two-spirit came out of that, of trying to have that identity outside of, you know, the prejudice from the Native community and the prejudice from the White gay-lesbian community or non-Indian gay-lesbian community.

The homogenizing effect of lesbian feminism, with its identity movement, positioned itself as "the expression of the aspirations of all women" (Stein, 1992: p. 558). The lesbian movement from the 1970s through the 1980s privileged White middle-class women for whom lesbianism represented a sexual object choice and political identity in opposition to white-male dominated systems. Lorde (1985) noted that lesbian women of color and white working-class lesbians were compelled by the privileged White middle-class majority to assimilate to their political agenda and identity politics, making this hegemonic lesbian identity their primary identity. This assimilation required a marginalization of race and class issues paramount in the lives of lesbian women of color and white working-class lesbians. As Roberta lamented:

How can one separate themselves from being two-spirit to being an Indian? So, I mean, to me it's a very hard concept to say that I am this or that. I'm all of that. Yes, I'm a Native woman, I'm two-spirit . . . it's all those things and encompassing and being, and for me to say I'm only this over here is not healthy because I'm not only this, I'm a multitude of things.

In response to the hegemonic identity politics of the White middle-class majority, many women of color challenged the notion of lesbian identity as organized around a fight against patriarchal influences and oppression and pushed for a more "diffuse notion of power and resistance" (Stein, 1992, p. 561). The women in our study were particularly uncomfortable with the anti-male separatism that is a core principle in some lesbian communities.

I have to thank the White gay and lesbian community for that because of their separatist issues. . . . I refuse to go into that little box,

shutting every [man] out, because they're an important part of all
creation and you know they're going to be there. (Sandy)

For Native women, embracing men was a way to fortify defenses
against White oppression.

A lot of times in the White community, lesbians will say, you
know, 'I just don't like men.' Actually, I think that identifying as
two-spirit, I have more of an alliance with Native men . . . because
they're Native men and they have experienced a lot of similar racist attitudes as well as homophobic attitudes on the reservation that
I have. Um, we seem to bond together better, the male and the female sides sort of complement one another. I have difficulty ex-
plaining it to White lesbians who would say, 'Well, why would
you want gay men at an event?' Because Native gay men are not
gay men, they're my two-spirit brothers. (Sandy)

Just as two-spirit women value women and men in the Native community, they embrace the feminine and masculine in their own two-
spirit identity. Wilson (1996) noted that the balance of feminine and
masculine qualities, of male and female spirits, embodied in persons is
often emphasized in traditional Native communities. Indeed, many Native origin stories speak to the balance of male and female, the importance of harmony between opposites as a way of wellness and
wholeness. For many of the two-spirit women, reconnecting with this
sense of wholeness meant also reconnecting with both the masculine
and feminine aspects of their spirit.

When we had pow wows at the two-spirit gatherings, there were
men who danced traditional women's dances and I think that's
definitely something that is a two-spirit thing that is important to
those men and like I haven't seen any women dance traditional
men's but, you know, when I was younger it was very important to
me to not–to be able to have that male side of my personality come
out and that was very repressed and I think especially younger
women really need to have that option of being able to have that
part of your personality. (Winona)

It's definitely a balance in understanding assets of both genders.
(Sandy)

It is no wonder that the term two-spirit emerged in the late 1980s, given the emergence of race and class-based ideological shifts to de-center the lesbian feminist model of identity during this time. For many of the two-spirit women, the lesbian feminist movement first served as an initial haven in which to share and bond with other lesbian women. However, it later became a community whose underlying ideology about men and gender relations failed to connect with indigenous women's realities and solidarity with Native men in the fight against colonialism and racist oppression.

Sadly, the term created by Natives for Natives, in their effort to secure their own identity free from oppressive influences, in part has been appropriated by the very forces they were struggling against. Consequently, the term *two-spirit* has acquired metaphoric power, becoming synonymous with spiritual power and ceremonial practices in an inaccurate and misappropriated fashion (Wilson, 1996). The metaphoric power associated with the term has led to romanticization and objectification of indigenous peoples who are two-spirit, in yet another example of colonial oppression.

> It's confusing sometimes because a lot of White people have started to identify as two-spirit and they don't get it, that it's not appropriate, you know. (Janis)

> I don't mean that, you know, that every gay Native person is, you know, a shaman or a guru, you know. That's the way White people want to see us . . . (Sandy)

"A UNIFYING NAME TO START TO HAVE A COMMUNITY": TWO-SPIRIT AS A COLLECTIVE IDENTITY TO SPUR SOCIAL MOBILIZATION

Two-spirit identity served not only to push away White dominance but to pull together the Native community with a collective identity in the struggle against racism, heterosexism, and internalized oppression. Many of the women described a two-spirit identity as a unifying construct that allows them to join with other Natives to explore their sexuality and gender from an indigenist perspective. As Winona remarked:

> People are acting like it [the two-spirit label] has been around forever, it has this deep meaning and all of this, which it has come to

have more meaning to different people but to me it was just something that was–something we needed at that time . . . it was just . . .
important to have kind of a unifying name to start to have a community of some common kind of things . . . we always knew everybody had their own traditions and communities we needed to go
back to, but as far as having support, it's not always realistic for
people to expect to be able to have support from other two-spirit
people in their own community . . . that name I think to unify people and to have a place where people can go and be safe and talk
about being two-spirit.

The increasing acceptance of the two-spirit identity among Natives
facilitated national and international gatherings that served to create a
safe space for identity exploration and development and to mobilize the
community.

I don't feel like I have to divide myself up so much anymore because I went to a lot of [two-spirit] gatherings in the early years . . .
so I have a lot of two-spirit friends . . . At the [two-spirit gathering]
there was a lake there and it was just nice, and we laughed . . . it
was a space where you could be normal, a week out of the year,
and so just that in itself I guess is what kept me going back to the
gatherings–even today where things are so much better, it's like
you just can't go anyplace and not feel like you have to protect
somebody or alter your identity or whatever. (Winona)

In general, the two-spirit gatherings every year are really helpful to
me because it feels like that's the only time in the year I get to be
my whole self in one place . . . to me it's healing. (Maxine)

The problem with a unifying term is that, although it is necessary for
political mobilization, it simultaneously privileges a single experience
of identity and diminishes within-group heterogeneity (Melucci, 1989).
As the Native women had observed in the White LGBT community,
unifying terms tend to privilege within-group dominant discourses and
marginalize other voices, a process antithetical to the original intent of
the term *two-spirit*. Indeed, social movement theorists have elaborated
upon the ways in which the U.S. political environment makes stable collective identities both necessary and damaging (Gamson, 1998).
    However, indigenous worldviews tend to embrace ambiguity, com-
plexity, and non-linearity–processes that run counter to group mobiliza-

tion for a singular unifying construct. Perhaps, then, this is why some of the two-spirit women noted the ambiguity in the construct of two-spirit. Instead of embracing a singular definition of the construct, they were comfortable with having it be a placeholder, a momentary construct that is readily contested and negotiated within Native communities and two-spirit spaces until a word is created that captures the fluidity of gender and sexual identity and the interconnectedness and inseparability of identity with spirituality and traditional worldviews.

> You know, even the word two-spirit as you may know is really just a contemporary marker . . . for lack of a yet-to-be-found better word. (Sandy)

> So even today, I probably would acknowledge myself more as two-spirit than anything else. Because it fits me better and it's not the name in our language that says it all, but it's probably about the best fit I can find right now and that's how I consider myself to be–two-spirit. (Roberta)

> I'm still kind of trying to figure out um, you know, what is the term for myself. (Janis)

"THEY COME OUT INTO THIS WORLD LIKE THAT": COMING OUT VS. BECOMING

From two-spirits' perspective, coming out to self and others might be better thought of as *becoming out* in the sense that this process of identity acquisition is really a process of becoming who they were meant to be–a process of coming home or coming in, as opposed to coming out or leaving an old identity behind to embrace a new one. Wilson (1996, p. 310) captured this experience: "We become self-actualized when we become what we've always been . . ." A two-spirit woman's assertion that she's "been this way from birth," conveys her understanding of how the Creator brought her into this world, this life, in a certain way that is directly connected to the ancestors and future generations in what might be called spiritual essentialism. This stands in contrast to the biological essentialism of mainstream lesbian discourses evoked by the same phrase, which alludes to genetic make-up or biological determinism. Hall's description of "warrior" women exemplifies this idea of spiritual essentialism.

They are just being, that is the way the Great Mystery made them.
They come out into this world like that. And they are living their
lives . . . they were just manifesting what they were. And how they
lived. It was something given to them by Spirit-this way of living.
(Hall, 1999, p. 274)

This process of becoming can be misread or missed by non-Natives, because two-spirits might not be out in ways that are consistent with
LGBT politics or anthropological paradigms. For example, the following incident illustrates how two-spirit behavior can be misinterpreted
when viewed through a Western lens.

When [an anthropologist] came to do her fieldwork, she said she
wanted to meet some "warrior women," so I told her, "O.K., come
to the reservation, I'll introduce you to some warrior women." But
[she] came back . . . and she said, "You know, they're just the kind
of women I am looking for but they do not know who they are."
Well, it is not that they do not know who they are, just because they
do not know the label anthropologists have put on them–because
they are just who they are. (Hall, 1999, p. 274).

"GO ALONG WITH THE PARTY LINE":
CONFRONTING COMPULSORY HOMONORMATIVITY

Just as two-spirit women struggled to create terms to describe their
identity free of White-dominated LGBT language, they also sought
ways of embodying their identity that were uniquely Native. Initially,
however, they often felt constrained by more rigid models of LGBT *appropriate* behavior. Winona described how she felt compelled by the
White-dominated lesbian community to be out.

You're expected to be out if you were going to be part of the lesbian
community. And if you weren't, it was like a bad thing and you
couldn't be accepted. And so the other thing was you're expected to
go along with the party line, which was very White oriented also. If
you didn't go along with that, you also were rejected. (Winona)

This compulsory homonormativity extended to LGBT rituals as well,
such as one's first *dyke* haircut. Although this may be a rite of passage
for some two-spirit women (e.g., Winona said, "[I] cut my hair off and
wore flannel shirts and all of that stuff") as well as mainstream lesbians,

the associated meaning often differs for two-spirits, as the following experience so poignantly illustrates.

> I cut my hair, proclaiming a new identity. . . . I know that in Cree tradition, we cut our hair when we are in mourning. When someone we are related to or someone we love dies, a part of ourselves dies. It is a personal ceremony. The hair, usually a braid, is buried in a quiet safe place where no people or animals can step on it or disturb it. There I was with a flattop, shaved on the sides and short, spiky and flat on top. My hair was everywhere on the floor of the flashy salon . . . people were stepping on it, walking through it, and eventually it just ended up in the garbage along with everyone else's. A connection with my community was buried in that garbage can. (Wilson, 1996, p. 312).

Assimilation into White lesbian culture quite often placed two-spirit women in the position of disconnecting from Native relatives and community.

> I really distanced myself from my family and um I felt like I had to for my own protection and I really got involved in the lesbian community, and, you know, we made other lesbians our family. (Maxine)

The cost of this disconnection, albeit temporary or transitional in nature, forced two-spirit women to split their identities, a process that can increase feelings of estrangement in both communities (Walters, 1997).

> I think when I was young and I moved into that house [with White lesbians], I guess I was really in denial about, you know, nobody knows I'm here, so I can just do what I want! And then I can go back to the Native community again and nobody will know. . . . When I did go back to the Native community, [I'd] go back in the closet again. (Winona)

> I guess I didn't have a clear identity, I kind of like divided myself up more, like I would hang out in the gay-lesbian community and just accept what the standard was there, and then I would be in the closet in the Native. (Winona)

In more extreme circumstances, this experience of disconnection led women to self-destructive patterns. However, women who discussed suicide and hopelessness found solace in their spirituality and in some cases, divine intervention, which helped them move through the most troubling parts of [be]coming out.

> I at 21, I almost committed suicide. . . . I remember putting the barrel of the rifle in my mouth. And I was getting ready to pull the trigger and all of a sudden I heard this chanting and singing and up over the hill came all these women and they were singing a song and they were chanting and they were circling around me and the sun came out so bright . . . that was such a profound experience . . . the women told me that there was importance in my life, that I was significant, that I was valuable, there were things in my life that I needed to do. And so to make it through the tough times I think about that moment. (Roberta)

Several women talked about how the effects of historical trauma and its aftermath in personal, familial, and community functioning also complicate the process of (be)coming out for two-spirit women.

> All Native people are dealing with trauma from our communities past and present. (Winona)

> When I came out, I had really a lot of confusion, and when I came out I didn't come out just as a two-spirit person, I was like trying to pull myself out of the muck and the dysfunction that had been my life. (Sandy)

> We still have a lot of homophobia in our own communities, there's still a lot of homophobia that I believe stems from the dysfunction in the heterosexual families in our communities . . . it takes a long time to unravel [what] you know that this society has forced down our throats over all of these generations, and it's definitely you know unraveling generation after generation of dysfunction and pain and it's gonna take a couple of generations before that finally, you know, is changed. . . . (Sandy)

"MORE OF A TRADITIONAL WAY OF DOING THINGS":
EMBODIMENT OF A TWO-SPIRIT IDENTITY

Many researchers have noted that cultural values are critical variables that shape LGBT people's self-concept, coming out processes, coping skills, and psychological well-being (Morales, 1989). For Native people, the effect of culture is powerful and pervasive, and many indigenous ethics directly contradict values held by LGBT communities (Walters, 1997).

Mainstream LGBT culture values coming out in an explicitly verbal manner, and there is empirical evidence that coming out in this fashion is linked to good mental health outcomes–at least among a White sample (Garnets & Kimmel, 1991). However, directness and drawing attention to oneself are not values within most Native communities, where cooperation, humility, and collectivism are more highly prized. Identity and authenticity are assessed by adherence to a traditionalist collective norm for behavior; accordingly, Natives often prefer more subtle, nonverbal ways of disclosing their sexual identity.

> I've always said I was a two-spirit person and I was able to do it in such a way that it wasn't like here I am, this is who I am, this type of thing, I'm going to be in your face with it. But in a mannerism of bringing it up and talking about it and being calm about it . . . I did it in a very non-threatening, a very non-judgmental way and that if I could sort of present myself with my identity and to be strong within that identity and to come from a spiritual way of life with it, it would have more of an impact than being a rebel radical would have. (Roberta)

> It's not really that I'm ashamed or that I don't want anybody to know because I mean everybody here knows. . . . It's more that uh in certain circumstances it's not helpful to sort of like push that in someone's face if they're, you know, like for instance when I . . . was at the elders' dinner, right. I didn't do anything that would indicate that part of my life because it's not appropriate, you know, it's an elders' dinner, you know. (Maxine)

> I think we focused on ourselves and people didn't come to a group and say, "We're here, we're queer, get used to us." You know? It was more people would hear about our group [of two-spirits] from other people and they would watch what the people in the group did. . . . I think people became more accepting because of that, because we didn't get up in their face about things and I don't know, I just think that happened back then at least with more people, that that's more of a traditional way of doing things. (Winona)

Native conceptualizations of time are not linear; the ancestors as well as future generations are viewed as omnipresent. Therefore, two-spirit people may not subscribe to the notion that they must quickly progress through successive phases to achieve a preordained end–in direct contrast to the sequence described by coming out stage models based on White LGBT experience (Falco, 1991). Journeys are more circular in nature; things come in their own time. As Debran, a Navajo two-spirit activist quoted in Farrer (1997), explained: "I think it is just going to take time. It takes time. And I have that time. I do not expect to be accepted just like that. Over time, over years, or whatever, as people see who I am and how I am, that is what it takes" (p. 300). Wilson (1996) explained how the indigenous value of waiting until the time is right allows individuals to examine their state of mind prior to taking an action. Ross (1992) reported that the less familiar the context, the more likely a Native person is to withdraw, be silent, and reflect on possible outcomes, given various courses of action as well as to be cognizant of whether the time is right for action. Sandy alluded to these values in discussing when to come out to others.

> I don't want to interrupt their process by saying, well, this is my experience. So now I find that I just sit back and I just kind of listen . . . there are some things that people need to learn in their own time and in their own way, and in their own space.

"SHE FULLY ACCEPTED ME": FAMILY AND COMMUNITY AS A SOURCE OF STRENGTH

Many of the women talked about a deep sense of belonging that could only be found for them in Native communities. Wilson (1996) explained that the support of family and culture plays a pivotal role in helping two-spirit women live authentically. She wrote: "Throughout my life, my family had acknowledged and accepted me without interference. . . . I acquired strength from my elders and leaders who were able to explain that as an indigenous woman who is also a lesbian, I needed to use the gifts of my difference wisely" (p. 313). Michael Red Earth, a male two-spirit activist, had a similarly supportive family: "For me, once I realized that my family was responding to me and interacting with me with respect and acceptance, and once I realized that this respect and acceptance was a legacy of our traditional Native past, I was empowered to present my whole self to the world and reassume the responsibilities of being a two-spirited person" (1997, p. 216). Roberta described how family and community support for her was personified in one relative, her grandmother:

> She fully accepted me for who I was and I was her granddaughter, and that's all that mattered. And it didn't matter what my lifestyle was, as long as I was happy. As long as I was healthy. And so I would say that my grandmother was a major turning point in my life, to fully be accepted by them helped me to continue and to look at life and to look at what is healthy.

The strength of familial and community bonds, however, was not always unconditional. Two-spirit women discussed their hesitancy to be open and visible to their own communities for fear of losing this important cultural support system and sense of belonging and being excluded from their primary supports in dealing with racism and colonialism. As Little Thunder, a Lakota two-spirit women activist quoted in Farrer (1997), remarked, "The pain of being rejected by one's own people can be the most devastating" (p. 311).

> I think when I say people just didn't have the option of being out before and some people still don't feel that, it's like you don't have this big huge community that you can go someplace and start a new life, it's like you just have this one small community and you can't afford to alienate yourself from that, so that has shaped a lot of the two-spirit type politics. (Winona)

Some women move away from their communities to live a more open life. Eventually, though, many return to their communities, willing to deal with homophobia in exchange for an otherwise safe space of shared understanding in the battle against racist oppression. Wilson (1996) described this trade-off as follows: "In Cree culture, 'silence' does not equal 'death' and to 'Act up' should not lead us to remove ourselves from our community. If it does, we seem most often to quietly find our way back home." Winona described her process of returning home as an act of compromise:

> It was really hard, but I was so sick of being in a really alien environment, of not having, even having my speech misunderstood and all of that. So I was kind of reacting to that, so the first couple of years weren't that bad, I was just so happy to be around some-

thing that was familiar again that I was willing to pay that price and after awhile it got to be a problem of–kind of that game where people know that you can't [come out], as long as you don't say anything and you put up with abusive crap from them once in awhile, that you'll still be accepted. (Winona)

"SQUAW!" "DYKE!" "BITCH!": *RACISM, HETEROSEXISM, AND SEXISM* *IN TWO-SPIRIT WOMEN'S EXPERIENCE*

Traumatic life stressors, in particular discriminatory events, loom large in the life of two-spirit women and have devastating consequences (Walters & Simoni, 2002). They are a contemporary manifestation of the succession of systematic assaults perpetrated by the United States government, including genocide; ethnocide (i.e., systematic destruction of life ways); forced removal and relocation; health-related experimentation; and forced removal and placement of Native children in boarding schools. These traumatic assaults are known among Native peoples as historical trauma and can lead to a *soul wound* or *spirit wounding* (Duran & Duran, 1999). Balsam and colleagues (2004) found that two-spirits in their urban sample had much more historical trauma in their families, childhood physical abuse, psychological symptoms, and mental health service utilization than their heterosexual Native peers. As Maxine remarked, "You know, there's just all the, the ways in which racism has shaped my life from my earliest memories." She reported experiencing discrimination and oppression in multiple forms, from verbal insults to direct physical assault.

I came out of the diner and these guys jumped me. And they were screaming at me, "Fucking bitch lesbian dyke!" You know, just, you know, and punching me and stuff like that.

Other forms of discrimination involve micro-aggressions–more insidious injuries such as the assumption that Natives are nonexistent or invisible.

People will come up to me and say, "Oh, I thought all Indians were dead." And you look at them and you think, and how am I supposed to respond to this? You know, Emily Post doesn't have an appropriate response for, I thought you were all dead! It's like, no we're not all dead! (Maxine)

So [the nightclub] was loaded with White women. I mean it was so thick, I just freaked–I panicked. It's like I felt I was drowning, because every time I said excuse me, no one would move. People saw me, but they wouldn't move out of the way–it's like I'm in THEIR space type of thing . . . it's like I'm always in the way, but, in my reality, they're always in the way [laughs], you know? (Roberta)

Other micro-aggressions involve objectification or romanticization of Natives.

I've had women come up behind me and grab my earrings and say, "You're Indian, aren't you?" And I want to look at them and say like, "What? What is your point? Yes, I'm Indian. Fuck off!" (Maxine)

Some micro-aggressions subject two-spirits to the role of educator, having to undo colonial and racist oppression for the unaware White individual. Sandy talked about her losing patience with this expectation.

Twenty years later, it's the same thing, and, you know, I think the larger organizations out there want to be inclusive but don't know how in a respectful way and . . . I'm just over that now and I don't want to be the one to try to teach them that because if they haven't gotten it by now, a part of me is like, well, what's the use? I've got other things to do. I've got a garden to plant. (Sandy)

Often two-spirits don't know why they have been targeted–which aspect of their being has offended the attacker.

You know the thing is that I still to this day, when people insult me or do things, you know, like if someone calls me a bitch, which happened to me on the [bus], when someone is rude to me, you know, I go through this thing in my head: Is it because I'm Indian? Is it because I'm poor? Is it because I'm a dyke? What is it that I did this time, you know, to, to have these people be hostile to me? . . . sometimes it's really hard to tell, you know, particularly when something is a verbal assault. I mean, you know, I've been called squaw so then I know clearly that that's about being Indian. I've been called witch and that's a little bit more, you know, what does that mean, right? And I've been called a dyke, well a fucking dyke

> [laughs] and a fucking squaw . . . don't know sometimes why I'm being attacked. I just know that I'm wrong to a lot of people. . . . I think there's a lot of hostility to the combination of things I am. (Maxine)

The women noted that even LGBT communities of color were not safe havens and were often oblivious to or unconcerned about Native-specific concerns.

> I have worked in different communities of color, been in women of color groups, but that also, a lot of times, I'm the only Native person there and here it's been mostly dominated by African American men and then for a while it was a group of Asian and African American women, and . . . just the prejudices between those two groups and the working out of how people are going to work together, and so I think it was good to work in that community but it didn't really touch my community that much . . . maybe I'd say three or four friends kind of made that effort to find out what's happening in the Native community. (Winona)

In Native communities, two-spirit women can elude racist oppression, but they often encounter heterosexism.

> [In Indian Country] it's safe to be homophobic. And in the cities, the urban [Indian] communities, I see a lot of violence, that is directed toward two-spirit people and sometimes it's hypocritical attitudes. "Oh, you're my friend and, yes, I really like you and I don't care that you're gay" and then that person walks away and then they [say], "God, you know, I don't want anybody seeing me with him because they're gonna think I'm gay, too." (Sandy)

The heterosexist attacks are sometimes veiled as another form of rejection.

> [Indian] people will use the pretext of something else to exclude you or if you were in a party environment, to beat you up. They were actually beating you up because they were homophobic, but it would be something else. . . . (Winona)

They might even occur during a ceremony, creating a particularly painful experience.

> I remember sitting there at a sweat lodge and the medicine was being passed around, one of the women saying, all of a suddenly she came out and said, "I would never drink from the same cup as a woman who was a lesbian." And it was just like–well, I hate to tell you this, but I just drank from it [laughter], you know? (Roberta)

There was some effort to contextualize the oppression from within Indian Country in terms of the historical trauma Natives communities have endured.

> I see us having problems of dissonance, from our families . . . you know, from each other sometimes because, unfortunately, some of us are so damaged we aren't trustworthy, and we hurt each other because we can get away with it. I mean that's one of the dangers of Indian culture is that because we are so oppressed, nobody gives a shit about what happens to us. So if we hurt another Indian person there are no consequences for that because traditional society has been so shattered for most of us, particularly living in an urban environment, there are no consequences if you hurt another native person. (Maxine)

Most of the two-spirit women felt Native communities had taken some strides toward combating heterosexism. However, with respect to the issue of racism in LGBT communities, they still had seen little to no progress.

> I think that once individual queer people started talking to elders in the Native community, they saw that they needed to look at homophobia, you know, and, you know, they've done some Cedar Circles up North and, you know, there's all kinds of different things happening in the Native community. And I don't see that happening in terms of racism at all . . . there's been some sort of pitiful little efforts here and there but it just hasn't, you know, anti-racism work has not progressed. Not far enough for me . . . in fact, I think it's in some ways it's worse and become more polarized. And there's a lot of attitudes like, oh, racism isn't a problem. (Maxine)

Winona talked about conserving her energy regarding doing anti-racist work in the mainstream LGBT community, because efforts in that area appeared futile.

> I don't spend that much time or energy in the White gay-lesbian community because I don't feel like there has been a big huge change in that community and so I don't really see a reason to put a lot of energy into that unless it's necessary, absolutely necessary.

Overall, there was a perception of significant change over the last few decades among Native people in terms of their attitudes toward LGBT issues.

> I think there are a lot of places where people aren't that accepting now, but it's an incredible difference between when I was in my 20's and now, between the acceptance of people . . . most [Indian] people feel like they should be somewhat accepting of gay-lesbian people and a lot of people really have worked on homophobia and are more accepting in the [Indian] community in general. (Winona)

Sandy displayed a spirit of hopefulness about the future, especially for two-spirit youth.

> You know, just little subtle, subtleties that happen . . . it's gonna take some time, but the strengths are that I see more and more role models for the younger people to follow . . . they have something to look ahead to. . . . (Sandy)

CONCLUSIONS

In this paper, we critically reassessed current theory on lesbian identity processes through the narratives of five American Indian two-spirit women. The narratives reflect dominant lesbian ideology from the margins, revealing overall how the women's experience of their sexuality and gender identity is inextricably linked to other aspects of their selves, particularly their spirituality and cultural heritage. The women described their identity and its development as complex, fluid, and emergent; they highlighted the importance of reclaiming and recreating two spirit roles; and they emphasized the need for countering oppressive dominant discourses. As they embrace their two-spirit selves, rather than *coming out*, these women found themselves (be)coming who they were meant to be and filling an integral space in their Native communities as caretakers, activists, and leaders.

The analysis demonstrated as well how indigenist forms of talk and identity construction are directly related to individual and community wellness. Because of their status as activists and leaders, the *testimonios* of these women and their actions are important for future generations of indigenous peoples. Their narratives weave a collective memory (Gongaware, 1999), not so much recreating a factual history of two-spirit experience but, rather, creating new norms and structures for social mobilization which can guide younger generations of two-spirits in their own journey–embracing ambiguity, contextuality, intersectionality, and above all, spirituality. Understanding the complex identities and roles of these women is key, then, to supporting not only the contemporary two-spirit community but for the well-being of future generations as well.

AUTHOR NOTE

This research was funded in part through a grant awarded to the first author by the National Institute of Mental Health (NIMH Grant R01 MH65871). Views expressed in this article do not necessarily represent those of the NIMH, National Institutes of Health, U.S. federal government, or University of Washington. We acknowledge the contributions of the HONOR Project's Leadership Circle of two-spirit elders as well as the national and regional Native community leaders, colleagues, and ancestors for their guidance and participation with this project.

REFERENCES

Anguksuar [LaFortune, R]. (1997). A postcolonial perspective on Western [mis]conceptions of the cosmos and the restoration of indigenous taxonomies. In S. E. Jacobs, W. Thomas & S. Lang (Eds.), *Two-spirit people: Native American gender identity, sexuality, and spirituality* (pp. 217-222). Chicago: University of Illinois Press.

Balsam K.F., Huang, B, Fieland, K.C., Simoni, J.M. & Walters K.L. (2004). Culture, trauma, and wellness: A comparison of heterosexual, lesbian, gay, bisexual, and two-spirit Native Americans. *Cultural Diversity and Ethnic Minority Psychology 10*(3), 287-301.

Bishop, R. (2005). Freeing ourselves from neocolonial domination in research: A Kaupapa Maori approach to creating knowledge. In N. Denzin and Y.S. Lincoln (Eds.), *The SAGE handbook of qualitative research, 3rd edition* (pp. 109-138). Thousand Oaks, CA: Sage.

Blackwood, E. (1997). Native American genders and sexualities: Beyond anthropological models and misrepresentations. In S. E. Jacobs, W. Thomas & S. Lang (Eds.), *Two-spirit people: Native American gender identity, sexuality, and spirituality* (pp. 284-294). Chicago: University of Illinois Press.

Brown, L. B. (1997). Women and men, not-men and not-women, lesbians and gays: American Indian gender style alternatives. *Journal of Gay & Lesbian Social Services, 6*(2), 5-20.

Falco, K. (1991). *Psychotherapy with lesbian clients: Theory into practice.* New York: Brunner/Mazel.

Farrer, C.A. (1997). Dealing with homophobia in everyday life. In S.E. Jacobs, W.T. Thomas, & S. Lang (Eds.), *Two-spirit people: Native American gender identity, sexuality, and spirituality* (pp. 297-317). Chicago: University of Illinois Press.

Gamson, J. (1998). Must identity movements self-destruct? A queer dilemma. In P.M. Nardi & B.E. Schneider (Eds.), *Social Perspectives in Lesbian and Gay Studies: A Reader* (pp. 589-604). New York: Routledge.

Garnets, L. & Kimmel, D. (1991). Lesbian and gay male dimensions in the psychological study of human diversity. In J. Goodchilds (Ed.), *Psychological perspectives on human diversity: Masters lecturers* (pp. 143-189). Washington, DC: American Psychological Association.

Gongaware, T.B. (2003). Collective memories and collective identities: Maintaining unity in Native American educational social movements. *Journal of Contemporary Ethnography 32*(5), 483-520.

Hall, C.M. (1997). You anthropologists make sure you get your words right. In S. E. Jacobs, W. Thomas & S. Lang (Eds.), *Two-spirit people: Native American gender identity, sexuality, and spirituality* (pp. 272-275). Chicago: University of Illinois Press.

Jacobs, S.E., Thomas, W., & Lang, S. Introduction. In S. E. Jacobs, W. Thomas & S. Lang (Eds.), *Two-spirit people: Native American gender identity, sexuality, and spirituality* (pp. 1-18). Chicago: University of Illinois Press.

Lang, S. (1998). *Men as women, women as men: Changing gender in Native American cultures* (J. L. Vantine, Trans.). Austin: University of Texas Press.

Little Crow, Wright, J. A., & Brown, L. A. (1997). Gender selection in two American Indian tribes. *Journal of Gay & Lesbian Social Services, 6*(2), 21-28.

Lorde, A. (1985). *I am your sister: Black women organizing across sexualities.* New York: Kitchen Table Women of Color Press.

McMahon, E. & Rogers, K.L. (Eds.) (1994). *Interactive oral history interviewing.* Hillsdale, NJ: Lawrence Erlbaum.

Melucci, A. (1989). *Nomads of the present: Social movements and individual needs in contemporary society.* Philadelphia: Temple University Press.

Morales, E. (1989). Ethnic minority families and minority gays and lesbians. *Marriage & Family Review, 14*, 217-239.

Red Earth, M. (1997). Traditional influences on a contemporary gay-identified Sisseton Dakota. In S. E. Jacobs, W. Thomas & S. Lang (Eds.), *Two-spirit people: Native American gender identity, sexuality, and spirituality* (pp. 210-216). Chicago: University of Illinois Press.

Ross, R. (1992). *Dancing with a ghost: Exploring Indian reality.* Markham, Ontario: Octopus.

Stein, A. (1992). Sisters and queers: The decentering of lesbian feminism. In P.M. Nardi & B.E. Schneider (Eds.), *Social Perspectives in Lesbian and Gay Studies: A Reader* (pp.553-563). New York: Routledge.

Thomas, W., & Jacobs, S. E. (1999). "... And we are still here": From *berdache* to two-spirit people. *American Indian Culture and Research Journal, 23*(2), 91-107.

Tinker, G. E. (1993). *Missionary conquest: The gospel and Native American cultural genocide.* Minneapolis: Fortress Press.

Tuhiwai Smith, L. (2005). On tricky ground: Researching the Native in the age of uncertainty. In N. Denzin and Y.S. Lincoln (Eds.), *The SAGE handbook of qualitative research, 3rd edition* (pp. 85-107). Thousand Oaks, CA: Sage.

Walters, K.L. (1997). Urban lesbian and gay American Indian identity: Implications for mental health service delivery. In L.B. Brown (Ed.), *Two spirit people: American Indian lesbian women and gay men* (pp. 43-65). Binghamton, NY: Haworth Press, Inc.

Walters, K.L. & Simoni, J.M. (2002). Reconceptualizing Native women's health: An "Indigenist" stress-coping model. *American Journal of Public Health 92*(4), 520-524.

Walters, K. L., Simoni, J. M., & Horwath, P. F. (2001). Sexual orientation bias experiences and service needs of gay, lesbian, bisexual, transgender, and two-spirited American Indians. *Journal of Gay & Lesbian Social Services, 13*, 133-149.

Wilson, A. (1996). How we find ourselves: Identity development and Two-Spirit people. *Harvard Educational Review 66*(2), 303-317.

Teaching Transgender in Women's Studies: Snarls and Strategies

Sara E. Cooper

Connor James Trebra

Sara E. Cooper, PhD, is Assistant Professor at California State University, Chico. She teaches courses on Contemporary Latina/o and Chicana/o cultural production, Modern Spanish American literature, and Gender Studies/Queer Studies. She has a special interest in women's experiences in Cuba. Her publications include one book, *The Ties That Bind: Questioning Family Dynamics and Family Discourse in Hispanic Literature and Film* (Lanham: University Press of America 2004) as well as critical articles in Chasqui, Confluencia, Dactylus, and several critical anthologies (most notably *Tortilleras: Hispanic and Latina Lesbian Expression* and *Reading the Family Dance: Family Systems Therapy and the Literary Study*). Dr. Cooper also has translated stories and novels by Latin American women writers. Her favorite activities include travel, reading, dancing, and playing with her niece, Eva.

Connor James Trebra recently completed his BA degree in Women's Studies with a minor in English at California State University, Chico. Some of his previous undergraduate academic work has included such diverse and eclectic topics as Queer Comix: Que(e)rying self representations in pictoral narratives, Queers in Russia: Trans issues and identities, and Christianity and Same-Sex Desire in Beowulf. He is the recipient of several grants and awards for Women's Studies students. Currently he is at work on an original project which will present a multidisciplinary approach to the study of female performance/identity in a paper entitled Women of the "West Wing": Visions of Beauty or Visions of Power. A longtime member of the lesbian/feminist community on both east and west coasts, he is currently undergoing physical transitioning from female-to-male, hoping to retain and enrich his identity as a queer educator.

Address correspondence to: Sara E. Cooper, Department of Foreign Languages & Literatures, California State University at Chico, Chico, CA 95929-0825 (E-mail: scooper@csuchico.edu).

[Haworth co-indexing entry note]: "Teaching Transgender in Women's Studies: Snarls and Strategies." Cooper, Sara E., and Connor James Trebra. Co-published simultaneously in *Journal of Lesbian Studies* (Harrington Park Press, an imprint of The Haworth Press, Inc.) Vol. 10, No. 1/2, 2006, pp. 151-180; and: *Challenging Lesbian Norms: Intersex, Transgender, Intersectional, and Queer Perspectives* (ed: Angela Pattatucci Aragón) Harrington Park Press, an imprint of The Haworth Press, Inc., 2006, pp. 151-180. Single or multiple copies of this article are available for a fee from The Haworth Document Delivery Service [1-800-HAWORTH, 9:00 a.m. - 5:00 p.m. (EST). E-mail address: docdelivery@haworthpress.com].

doi:10.1300/J155v10n01_08

SUMMARY. We moved from the San Francisco Bay Area to a small rural town in Northern California, thinking that with our years of experience in the queer and trans community we would go out to change the world. Instead, teaching Women's Studies students about transgender at Chico State changed us in ways that we never would have predicted. *[Article copies available for a fee from The Haworth Document Delivery Service: 1-800-HAWORTH. E-mail address: <docdelivery@haworthpress.com> Website: <http://www.HaworthPress.com> © 2006 by The Haworth Press, Inc. All rights reserved.]*

KEYWORDS. Teaching transgender, transgender education, GLBT education, transgender feminism, transgender theory, women's studies, gender studies

Thinking of all the many ways that the word, the concept, the action *transgender* impacts my life, I realize that they all come together (strangely enough) in the classroom. Once a year officially, and numerous times throughout the year in practice, I teach undergraduate students in a rural state university about the transgender continuum. This task is complicated by a plethora of factors, including my own complex sexual orientation, the students' relative lack of life experience and surfeit of prejudices, the tendency of some Women's Studies courses to emphasize a gender dichotomy and a second wave feminist viewpoint, and of course the inherent perplexities of the trans issue itself. Yet this is one of the course topics that is most personally fulfilling for me and perhaps the most enlightening for my students, once we can surmount the hurdles that block our path. This has sustained my commitment to engage in reflection about how best to face and deal directly with these pedagogical obstacles rather than simply to circumvent them. In this way I can continue to support "queer persons and citizens [who] are still immersed in the difficult battle for mainstream presence and place" in the sphere of education (Grace and Hill, paragraph 11). *From my perspective as well, teaching transgender to college students, who may or may not have had previous exposure to even gay or lesbian individuals, is both challenging and rewarding. It is also necessary if we as educators are to instruct and inform a new generation regarding the viability of alternatives to the proscriptive and prescriptive dualities of a bigendered society.*

When I read the call for papers for a special *Journal of Lesbian Studies* volume on Intersex, Transgender and Queers in Lesbian Communities, I began to ponder once more just how inescapably my teaching of transgender is connected to the sometimes tense and volatile relationship between the transgender movement (or indeed single transgender individuals) and the lesbian and feminist communities.[1] I wanted to explore how the lesbian/transgender friction has been problematic in the classroom, and how my own life has been impacted by the emerging tensions. So I brought up the idea with Connor. *At first resistant, unsure of the context in which our paper about "teaching transgender" would be presented, as well as wondering about the way my own contribution might be received in a journal with "lesbian" in the title, I nonetheless became enthusiastic and hopeful that this contribution might not only aid educators in their attempts to teach the "T," but that it might also contribute to a more reasonably dialogic atmosphere between transqueers and lesbians–a relationship that has often been strained and confrontational.* So we were on, and this paper is the result of our collaboration; more than anything else, our work for this project has allowed us to study in depth and articulate some of the ideas that had been simmering beneath the surface. The main objective of this essay is to explore the transformative power of teaching the transgender continuum within the context of the Women's Studies program at the California State University at Chico (CSUC). Nonetheless, I do not want to elide the actual challenges of teaching trans, so I will review some classroom incidences over the last three years and discuss my experience with particular texts and exercises. My partner of seven years, Connor James Trebra, also will reflect upon his experiences as a guest lecturer on transgender in my classroom. Finally, we will discuss how teaching transgender together has changed our lives radically.

A brief explanation of my own position is in order.[2] I am a professor of Spanish and Women's Studies, hired principally to teach the former, but "lent" to the Multicultural and Gender Studies program to teach the fledgling course entitled: LGBTQ: Issues and Identities. As a femme, I could easily have passed as heterosexual on the job market, but chose instead to be vocal about my "same-sex domestic partner"; as a graduate of a traditional program in Spanish, I could have passed as a specialist in women writers, yet I preferred to be clear about my interest in queer studies. There is a certain tension involved in my visibility, compounded by the strain of resisting the urge to exist under the radar. Fortunately, my fascination with narrative and performance helps to alleviate or in other ways mitigate the pressures related to living a queer identity in the academic world. I am able to embrace a fluid expression of self, of womanhood, of sexuality as is consistent with my changing self-perception and/or the circumstances in which I find myself; thus, every encounter within or without the classroom is a potential adventure, a teaching and learning opportunity in the making.

I am told that my hire was controversial, because some feared that my "eclectic interests" would take away from my dedication to the Spanish program. Despite possible hidden messages of homophobia in this rumor, and although I am not the only out lesbian/gay faculty in my institution (by far), I have been welcomed as a sort of lesbian mascot at my university. I have been invited to give numerous guest lectures on lesbian and gay topics (in classes of Sociology, Philosophy, and Women's studies); I was given the LGBTQ course; I have been asked to mentor students, participate in events, and give public lectures that highlight my sexuality. Perhaps my particular brand of intellectual diplomacy and my fairly traditional feminine appearance (somewhere between professional and bohemian) has been nonthreatening and easy to swallow. Even my ultra-butch partner was welcomed into the university community and is now pursuing a degree in Women's Studies. Truth be known, I think that both my community and I have found a comfort zone in allowing Connor's "demonstrably defiant deviation" (Martin 106) to stand symbolically for my own transgression. I am free to perform any extreme of femininity without my sexuality being in question, as long as my image is countered by that of my partner (whether explicitly seen, remembered, or even imagined). As his course of hormonal therapy progresses, we speculate endlessly on how colleagues and community will react to this new twist. Will they accept us as even closer to the norm, ceding us elements of heterosexual privilege, or will this put a mile-high kink into the calm of our relations? Will my degree of credibility as an activist and a professor of queer studies be threatened if others interpret our physical appearance and linguistic choices as a wish to pass?

I could understand some interruption of smooth sailing, on the basis that the feminist academic community shares so many of the biases that I have seen in the lesbian communities of which I have been a part-first in Austin, Texas, then in the California Bay Area and now in the northern California town of Chico. I have heard the arguments about what constitutes a real woman (read: who would be accepted into womyn's space) and on more than one occasion I have seen a group of women drop an FTM like a hot potato as soon as he came out of the closet as more (or less?) than a butch lesbian. As FTMs come out of the closet a

second time into new and different spaces of queer identification and resistance their struggles and challenges may remain unseen, or perhaps
perceived as moving into a conformist and patriarchal freedom.[3] To my
horror, I have even heard staunch feminists reverting to blaming the victim when bisexuals or transgender individuals were raped. Let us not
forget–this is only three decades from the lesbian feminists who in the
1970s placed a series of rules on sex between women including prohibitions against penetration, role-playing, use of erotica, or even one
woman lying on top of the other (*possibly* forgiven if both partners were
scrupulous about spending equal time in the physically top position).[4]
Truly, for a marginalized group subjected to harassment, oppression,
and pressure to conform from the "sexual majority," not to mention the
patriarchal system of gender discrimination, we lesbians have an awfully bad reputation of defining what is authentically female or lesbian,
restricting access to the small range of rights and privileges won by the
gay and lesbian movement, and making negative judgments on nonconformity. Therefore, it is hardly surprising that I have seen so many
raised eyebrows, amused chuckles, and looks of shock or disgust over
my inclusion of the "Bisexual-Transgender-Queer" end of the spectrum
in public speeches, informal conversations, and classroom lectures. Yet
as you can see, even if I was not absolutely convinced at the pedagogical
necessity of teaching the full range of queer identities, I would still feel
the need to educate students, faculty, staff, and community members
partly as a matter of personal safety and respect.

THE COURSE:
LGBTQ ISSUES AND IDENTITIES

LGBTQ Issues and Identities is described in the online university cata-
log as "An exploration of current scholarship in gay, lesbian, bisexual,
transgender, and queer theories, issues, and communities. Grounded in
feminist scholarship, the course examines LGBTQ identity construction and formation through media, politics, sex/sexuality, science, and
the law. Specific focus will vary from semester to semester." The listed
course objective is to be able to thoughtfully discuss and analyze
LGBTQ communities and concerns in terms of:

- Theories–Scientific, religious, social and literary–that people have
  developed to explain sex/gender.

- Ways these theories and concerns have been developed, revised, contested, and enacted over history and across cultures.
- Ways in which sex/gender constitute biological, cultural, economic, and psychological forces.
- Ways in which sex/gender–and conceptualization of sex/gender–have affected people's lives, from the personal to the political.
- Ways in which sex/gender is perceived distinctly according to the racial, ethnic, cultural, socioeconomic, and religious context.
- How LGBTQ issues and identities impact creativity, arts, values, and reasoning.
- How these issues and identities relate to the students' own lives and community (WMST Syllabus, 2004).

The curriculum has varied somewhat over the past three years, but always has included several foundational elements. The first couple of weeks present historical background on same-sex desire and the invention of the concept of homosexuality. We then spend a week on each of the main identity categories, Gay, Lesbian, Bisexual, and Transgender. A week is dedicated to Queer Theory. The rest of the semester provides in-depth studies on a variety of specific issues and/or genres of cultural production, such as religion, family, AIDS, film studies, animation, fiction, and poetry. Attention is paid to intersections with race and class throughout the entire course. In creating and implementing this curriculum, the Multicultural and Gender Studies program hoped to add balance to the Women's Studies major, to provide support for LGBTQ minority students, and to enrich the education of a wide cross-section of the CSUC population. According to Tetreault's Feminist Phase Theory, adapted to reflect inclusion of LGBTQ discourse, the addition of this course brings Chico State somewhere between Phase 2 and Phase 3. Currently faculty are working on a proposal for a minor in Sexual Diversity Studies, which would raise Chico State to Phase 4 of integration of LGBTQ experience into the curriculum.[5]

Of the three years that this course has been taught, fall of 2001 enjoyed the largest enrollment and the greatest diversity of students. The sixteen students (fifteen enrolled and one auditor) were comprised of five men and eleven women; four students identified as persons of color, and the rest identified as white. Students' ages ranged from 18 to 35. By the end of the semester, two students had self-reported as heterosexual (one male and one female), seven as gay or lesbian, and the remaining seven students as bisexual, fluid in their sexuality, or undecided. In this last category, two suggested that the category of transgender might provide the best description of their identity. The second year the course was taught, fall of 2002, only six students were matriculated, one man and five women, all of whom identified as white. The age range was similar to the first semester's class. One student identified as heterosexual, two as lesbians, and three as queer or fluid in their sexuality. None felt that the transgender category fit. The most recent configuration of the course, from fall 2003, included nine students, all of whom were women and none of whom identified as transgender. All of the students were under thirty years old, most being in their early to mid-twenties. Six students self-reported as white, and the other three as non-white. Throughout the course, two students described themselves as heterosexual, one as lesbian, four as bisexual or fluid, and two students did not state their sexual orientation or preference.

THE TRANSGENDER UNIT: TEACHING RESOURCES AND STRATEGIES

My students often find it challenging just to grasp the fact that gender is not as simple as the ubiquitous two boxes have led them to believe; although our campus is proactive in fighting for women's rights and is learning to be so with gay and lesbian rights, the activism is firmly rooted in gender difference.[6] Next, we are faced with the prickly problem of trying to understand FTMs and MTFs within various lesbian feminist contexts and concepts, including: pride in the biological female body, women-centered and women-loving-women exclusivity, and the fight to end compulsory heterosexuality. In terms of Tetreault's Feminist Phase Theory, adapted to evaluate the inclusion of LGBTQ discourse, most of the curriculum at Chico State–thus the students' preparation–is at Phase 2 (Compensatory Scholarship) or Phase 3 (Bifocal Scholarship). Lamentably, many courses still are at Phase 1, representing heteronormativity as the universal experience. My colleagues and I who are at Phase 4 (LGBTQ Scholarship) or 5 (Multifocal, Relational Scholarship) are thus faced with several challenges specific to working in the classroom with students who have not yet developed a similar critical apparatus.[7] My pedagogical goal is to uphold–and expand–the students' sense of self-empowerment and respect for all people while challenging the prejudices that a genderist and womanist perspective can instill. I purposefully have chosen texts and exercises that directly address the problem. Understandably, upon reading or hearing testimonials by trans warriors, Women's Studies students are aghast to learn that transgendered folk have become the new pariahs in the lesbian community. *This most obviously may stem from genuinely empathetic feelings of outrage, or perhaps from subconscious anxiety that even the "queerest" of communities will not accept the transgendered, the result of which may be that they must now blend in with normative society.* However, students in this course have worked hard to comprehend both the definition of transgender (the lack of congruence between one's body and one's self-perception, often complicated by impossible or repugnant societal expectations) and the dilemma of the transgender community. The students understand that not only do many transgenders mature, come out of the closet, and come to a greater awareness of themselves within the gay and lesbian community, but also they continue to resist heteronormativity and share a particular sense of themselves as queer in spite of going through levels of physical transition. Transgenders face discrimination, bigotry, harassment, and abuse like gays do. When Women's Studies students realize the intrinsic similarities between homosexuality and transgenderism, many are ready and willing to become activists for transgender rights.

Before taking this class, many students are under the mistaken impression that the GLBTQ community is and always has been cohesive and mutually supportive of all its constituent parts. They assume that the only real threats and misunderstandings come from the conservative heterosexual sphere. To correct this assumption, we take an historical approach to studying the various identities and communities as they emerge and differentiate themselves, with all of the concomitant fears and struggles between factions. The Gay and Lesbian movement arose in opposition to the assimilationist Homophile agenda; lesbian separatists worked to counteract gay male control as well as homophobic elements among the feminists; bisexuals and pansexuals have yet to be accepted as individuals who already have a clear orientation rather than fence-sitters who have not yet made a decision (or admitted the truth) about their sexuality.[8] As each oppressed subgroup begins to win some degree of acceptance and civil rights for themselves, it acknowledges an acceptable range of identity and establishes a set of parameters that constitute acceptable behavior, or in other words, a new "norm." As gays and lesbians claw their way into a safer and more solidified space in society, they experience justifiable fear and anger toward an ever more complicated identity politic that undermines the solidity of gender or sexual difference, which defines their identity, or that seemingly puts into question the respectability of gays. By the time the class delves into the topic of transgender, we are accustomed to the idea that even in the

GLBTQ community the dominant majority will attempt to control or suppress "alternative" identities that deviate from the new "norm." Yet the extent of the bellicosity exhibited between the transgender community and much of lesbian society still shocks and appalls us.

In an introductory course on GLBTQ, it is impossible to fully study the transgender continuum in all of its complexity. For the sake of brevity and comprehensibility (and knowing that any set of definitions will be somewhat arbitrary and fail to satisfy), I explain transgender as an umbrella term that could include transgender warriors,[9] gender outlaws,[10] two spirit, tomboys, sissies, FTMs, MTFs, transvestites, transexuals, and intersex–and more. I do recognize that it also is necessary to discuss the fact that individuals, and even activist organizations, associated with the aforementioned terms resist inclusion in the transgender category. Before anything else, I want the students to look beyond the boxes that have framed their existence, whether those boxes are in their own mind or on the application for a driver's license, in order to understand the inescapable insufficiency of the two-gender fallacy. Most years I have set this up from the beginning of the semester, when students complete the "Are You Perfectly Gendered" self-test from Kate Bornstein's *My Gender Workbook* (46-62). Secondly, I want them to begin to fathom the importance of self-definition of gender as well as external definitions that originate both from gender activists and from sympathetic psychological and medical professionals. In my view, these definitions would be qualitatively different from those proposed by inimical individuals and organizations, including hostile doctors, psychiatrists, gays and lesbians, and feminists. Students do need to understand the depth and breadth of the opposition to transgender (in theory and practice), especially the historical intolerance shown by the lesbian (and lesbian feminist) community. Nonetheless, the end result of this unit is that students see the strength of transgender resistance and the legitimacy of the fight for transgender rights. Ideally, since the Women's Studies program at Chico State is activist in nature, students will embrace some degree of pro-transgender activism, from educating their friends and family to joining the battle for social/legal equality.

To achieve these objectives, I have utilized a variety of class activities, grounded in a diverse body of readings–from sociological studies to personal narratives to popular culture texts or films. In this course, reading personal narratives of transgenders alongside erudite scholarly essays augments students' understandings of each; the confessional tone and vulnerability of the testimonies pushes students to relate on a personal level and access levels of comprehension that are not engaged by intellectual arguments. Many students take this course out of interest rather than for credit, so the appeal to their emotions and experiences is particularly effective. On the other hand, students need to realize that queer studies in general and transgender studies in particular are credible and fruitful areas of research. David Román warns the instructor "it is essential to this process [of pedagogical growth], of course, for us to develop methods of interrogating the various theories, scripts, and performances of sexuality that problematize a lesbian or gay identity" (113). A decade after Román's statement, we can see that this is doubly true when teaching the transgender continuum. Moreover, students certainly recognize when they are not challenged to use their critical faculties to expand their knowledge base or to engage in critical scrutiny of ideas (especially true in our Women's Studies program), so the scholarly treatises balance out the readings, providing tools and models of analysis. In response to students' positive reception of the texts they read in the first semester the course was given, core readings for the transgender unit have been fairly uniform over the last four years. Students consistently have read "Living Our True Spirit" from Leslie Feinberg's *Transliberation* and "Nuts and Bolts" from Kate Bornstein's *Gender Outlaw*, in addition to Serena Nanda's essay "Multiple Genders Among North American Indians" (in *Gender Diversity: Cross-Cultural Variations*) and Beth Brant's short story "Coyote Learns a New Trick" (in *Mohawk Trail*). After the first semester I dropped a chapter from *Boys Like Her: Transfictions* and added Judith Halberstam's rather dense article "Transgender Butch: Butch/FTM Border Wars and the Masculine Continuum" from *Female Masculinity*. From Brett Beemyn and Mickey Eliason's critical anthology entitled *Queer Studies*, a student group will report on Ki Namaste's "'Tragic Misreadings': Queer Theory's Erasure of Transgender Subjectivity." Encouraged by the spirited conversation that transpired in the pilot year of the course, I have always maintained space for at least an hour's discussion of these readings. "Nuts and Bolts" offers humorous and explicit (albeit exceedingly brief) comments on the MTF surgical process, hormone therapy, and sexual satisfaction afterward. The writing is engaging and is short enough to open the discussion, yet it leaves many questions unasked and unanswered–the pedagogical advantage of which is priceless in this context. This effectively sparks the students' curiosity; they begin to proffer direct questions about body parts, surgical processes, costs, psychological requirements, and other specific elements introduced by Bornstein. The specificity and technical language of the article provide a model for open and relatively unashamed dialogue about transsexuality; every year students ask about both the particulars of the surgery and the various motivations behind choosing a full transition. The discussions tend to be fluid yet wide-ranging explorations, which of course mirrors the elusive quality of the topic under study. Often the students come to understand the slippery nature of transgender identity and the impossibility of pinning down one clear definition for it through participation in such a conversation.

The Namaste and Halberstam articles introduce vitriolic controversies around transgender, both in the lesbian community and in the academic sphere of queer studies. Although both pieces are jargon-rich and somewhat difficult to wade through, students have appreciated the frank criticism and intelligently phrased arguments that elucidate and condemn the poor treatment of the transgendered within the GLBTQ community. From my observation, these two essays contribute to the students' grasp of the situation, their recognition of key terminology, and their ability to talk intelligently about problems with feminists' and gays/lesbians' defense of the gender dichotomy. Feinberg's contribution is the transcript of a keynote address spoken at a True Spirit Conference (in Laurel, Maryland), where all sectors of the trans community, including allies, are called to unity. This inspirational speech has been hands down the favorite reading of the unit every year. Admitting to personal experiences of oppression and harassment at many levels and from many sectors, Feinberg in no way sounds naïve or trite. S/he acknowledges the rage and frustration that have plagued her/him, but convincingly argues that mutual forgiveness and building unity in diversity (strength in numbers) are necessary to achieve transgender liberation. *This is consistent with Feinberg's socialist-Marxist reading of political challenges facing marginalized communities (sexuality, racial, and socioeconomical).* Feinberg's message of cautious optimism echoes the desires that the students often express throughout the course: that the GLBTQ community refrain from infighting, that alliances be forged across the factions and with open-minded heterosexuals, and that the focus be turned toward human rights and social justice for all. Students often become quite animated and inspired in the discussion of Feinberg and respond with enthusiasm to a class activity brainstorming individual activism they can employ on a daily basis. In essence, the mixture of readings (personal narrative, fiction, and scholarly texts) and activities allow a more dialogic interaction in the course, whereby students combine reflection and action.[11]

Nanda's essay and Brant's short story are readings from later in the semester and as such reintroduce the topic of transgender from a distinct angle. "Multiple Genders Among North American Indians" reinforces several key concepts, such as the fact that perceptions of gender and sexuality are culturally and historically specific. Students are intrigued by the idea that hundreds of years ago indigenous civilizations of the Americas sustained gender systems that allowed for gender reassignment based on an individual's interests, skills, and/or desires (rather than anatomy). They are again challenged to comprehend and accept cosmologies that designate three, four, or more distinct genders. Perhaps the most astonishing element for the younger students is that the gender continuum was not originally or exclusively discovered (or invented) by someone in the United States, or even Western Europe. Even the students of color who have taken other Multicultural and Gender Studies courses tend to express surprise, as contemporary Native American culture is often portrayed as homophobic and misogynist. Few are nonplussed, however, by the suggestion that less flexibility was allowed to individuals born as female; somehow this detail makes the entire premise easier to swallow, especially for the Women's Studies students. Read alongside of the scholarly essay, Brant's story generates enthusiasm, laughter and delight. This transgender refashioning of a classic trickster tale is the odd man out, so to say, in a collection of stories about crones, wives, mothers, and daughters. Upon being asked, students eagerly have speculated about the reception of the Coyote story by the Native American and feminist audiences that the book undoubtedly attracts. More than one especially observant student also has remarked that the story could be problematic for a transgender reader, being that the cross-dressing and chest-binding trickster in essence is told to stop pretending and start enjoying a real woman-to-woman sexual encounter at the end of the tale. All of the core readings have been successful in sparking lively debate and in bridging the theoretical and emotional gap in knowledge.

On the other hand, the overall class format and scope of activities have changed each year. Before or during class, students have viewed fragments of such diverse feature and documentary films as *Adventures of Priscilla, Queen of the Desert* (1994), *Different for Girls* (1997), *Southern Comfort* (2001), *Boys Don't Cry* (1999), *Gender Unknown* (2001), *Is It a Boy or a Girl?* (2001), and *Normal* (2003). At times, film has been used merely as visual reinforcement, other times to introduce the authoritative voice of a transgender individual or a medical professional, and other times to add an emotional or personal element to what can be an overwhelmingly esoteric and complex discussion. In the first semester, a panel of transgender, butch and femme women spoke to the class; in subsequent semesters panels have included only one or two individuals who did not identify as their assigned gender. However, in the second and third years of the course, a student from the first semester returned as a guest lecturer, speaking from years of experience in the San Francisco Bay Area LGBTQ community and a growing level of academic expertise in transgender. This happens to be my spouse, Connor James Trebra, who will share some of his experiences as guest lecturer in the next section of this essay. Even with the continuity of Connor's inclusion, or perhaps as a direct result of the same, our approach to teaching the topic of transgender persists in evolving. One year the focus was much more theoretical and clinical, which was less satisfactory overall. Last year the guest lecture was considerably more personal, and at the same time showed a firm basis in current scholarship. This combination, along with a call to activism, was very well received by the students.

As the number of individuals who are willing to identify publicly as transgender(ed) multiplies, encompassing a spectrum of social and professional positions, the literature, whether academic or narrative/biographical, also increases. In order to provide students with a wide range of information, it is essential that the material used in courses that deals with GLBT issues/identities be kept current and contemporary. Providing historical background for the multitudinous queer identities, how they have been viewed and shaped, is important; but as we begin the 21st century, educators, researchers and students must be informed of all the different ways in which the transgender community is interwoven into the fabric of society. As the political climate in the United States becomes more repressive and conservative, as targeted citizens' basics rights are being challenged by laws passed in the name of patriotism, as marginalized populations are being denied access to health care, housing, and education, and as traditional values threatening the rights of gays and lesbians to marry, adopt, and hold mutual property intrude into the legal and cultural aspects of "American" society, it is imperative that we provide students with the intellectual tools that will enable them to make informed, anti-homophobic/transphobic decisions for themselves and for their children. Keeping our curriculum current enables students not only to identify, but to see the need(s) for positive change as essential to the future. Some of the additional works that could be particularly useful for a general background or student research projects are Aaron H. Devor's extensive and academic tome FTM: Female-To-Male Transexuals in Society; *Jason Cromwell's more trendy and casual* Transmen & FTMs: Identities, Bodies, Genders &

Sexualities; *Henry Rubin's thoughtful and politically relevant sociological study of urban and coastal* FTMs Self-Made Men, Identity and Embodiment Among Transsexual Men; *and Riki Anne Wilchins' provocative* Read My Lips: Sexual Subversion and the End of Gender. *Two helpful books from the healthcare professional sector, both of which list valuable resources, include* True Selves: Understanding Transexualism (For Families, Friends, Coworkers, and Helping Professionals) *by Mildred L. Brown and Chloe Ann Rounsley, and* Transgender Care: Recommended Guidelines, Practical Information & Personal Accounts *by Gianna E. Israel and Donald E. Tarver II, MD.*

TESTIMONIAL:
TEACHING FROM A TRANSGENDER MIND AND BODY

As a female-bodied transgendered individual who is presently undergoing hormonal therapy to physically alter the body in which I have lived for many years, I have personal experience in the area of "teaching transgender" in classroom settings–as a member of GLBT panels, as a student who is very out among his peers, and as a "guest" lecturer. The narrative portion of my contribution will involve thoughts and experiences gleaned from presenting and leading class discussion about transgender as a student-classmate, and from my experience of lecturing once a year to a Women's Studies class (GLBTQ: Issues and Identities) on the "trans" issue at CSU-Chico.

After spending thirty years living in the relative safety of the queer ghettos of the San Francisco Bay Area, I chose to move with my lesbian lover to a small college town in Northern California. As mentioned above, she is very femme and some are surprised when she announces that she is not heterosexual. On the other hand, people are never surprised at, nor have I ever had to announce, my non-normative sexual identity or gender presentation. We immediately established ourselves as the newest queer couple on the block and within a year were involved in many layers of both the academic and queer communities. I attended nearly all social events sponsored by my spouse's department and the university. We attended the local lesbian potluck socials and a weekly dinner with a dozen heterosexual members of Sara's faculty cohort, became board members of the local gay/lesbian organization, and enjoyed university concerts and lectures. I was introduced to my spouse's colleagues, as well as to her chair, her dean and the president of the university, as "my partner, CJ." I wore the suits and ties; she wore the dresses and stockings. Surprisingly, for the most part we both were treated well, and many colleagues extended their friendship and support. During our first semester at Chico, I audited several classes–Spanish, anthropology, Russian history, and women's studies–and always put my own special queer bent on papers and presentations. Students in my classes were either polite or barely saw me–I was an older student, over 50, and perceived as a dyke lesbian, and in that permutation experienced the invisibility to which any female-bodied person of that age is subject.

In fall of 2003, I became a full time student, majoring in Women's Studies with a minor in English. In January 2004, the beginning of my second semester at the university, I began my FTM transition in full view of students, faculty, friends and family. How a 55 year old female-bodied human, after spending nearly 35 of those years as a lesbian-feminist, makes the decision to alter his body to better reflect his gender identity, is another story, one that will not be told here. Suffice it to say that making the decision literally took years, a great deal of reflection and, as usual, the support of a very fine woman. However, some biographical information is pertinent and necessary for my contribution herein.

Because of my age, I have been situated at the front lines of the burgeoning second and third wave of feminism. Starting in the very early 1970s, I attended lesbian/feminist conferences, consciousness raising groups, women's music events, and marches. During that time period, I was present at a small conference in Limmerick, Pennsylvania, for a weekend "retreat" which featured Rita Mae Brown and other members of the newly formed Radicalesbians; I marched in New York City in front of the Tombs to protest the treatment of women in prison; I was part of a group of women in New Haven, Connecticut, who attended the trial of Ericka Huggins and Bobby Seale, collaborating on a broadside publication that was sent out to various individuals and radical groups throughout the country; I was present in San Francisco for some of the first discussions regarding women, class and race; I have marched for AIDS research and was present in San Francisco during the assassination of Harvey Milk and George Moscone, and participated in the subsequent White Night riots. I have participated in, as well as been supportive of, queer and women's struggles, functions and establishments throughout my life. My primary relationships have always been with women, and although I have often felt like a fraud, I am most comfortable in the presence of female energy. These experiences combine to afford me the situatedness of a woman-centered consciousness and female socialization.

The above experiences, along with the avid reading of feminist/lesbian analysis and interpretation, afford me a position from which I may critically approach many subjects. The teaching of transgender, however, is much more complicated than an in depth reading of de Beauvoir, Halberstam or even Butler. It is certainly more complex than teaching English grammar, Russian history or even the history of the GLB movements. From my perspective, transgender issues are far removed from objectivity–they are rather steeped in profound subjectivity. Therefore, creating and presenting an overview of transgender issues and identities, within a women's studies environment and with a feminist perspective, to students actively participating in women's studies courses has been challenging to say the least. It often involves the delicate balance of compulsive disclosure and presentation of a variety of other subjective interpretations and self-representations. Moreover, despite recent advances in trans-centered research and writing, the available discourse, both narrative and academic, is very limited, sometimes contradictory, and often clinical and pathologizing. Transgendered academics represent both social constructionist as well as biological essentialist points of view. These discourses are often criticized and dismissed by others who ardently defend their own subjectivities and frequently present students with political and cultural interpretations which themselves are hegemonic, concretizing alternative pre and proscriptions.

In my first semester of WMST 110, my presence was both voluntary and requested. Because of the above experiences, my partner felt that my attendance in this new class would be invaluable to a roomful of 18 to 22 year olds–I had, after all, lived much of the recent history we would be discussing. Therefore in the fall of 2001 I audited this LGBT course. I participated fully in the class, adding my older and somewhat "queerer" inclination to many of the topics under discussion. In the fall of 2001, during which time my partner attended an out-of-town conference, I was asked to lead discussion on the film Boys Don't Cry *and on trans issues, including the definitions of the sometimes confusing terms of transvestite, transgender, and transsexual and their relationship to sex, sexuality, and gender identity. I did so from a personal subjective perspective. This lecture was repeated in 2002 and again in 2003; each subsequent presentation became longer and more extensive, fleshed out with film clips, definitions from both AMA and transgendered voices alike, as well as excerpts from narratives and conference addresses.*

The discussions which evolved from these presentations ranged from mild resistance, to dogmatic rejections, to wide-eyed disbelief regarding the hatred and discrimination that trans people must deal with on a daily basis, to admiration for the courageous and persistent way transfolk live their lives. All discussions were thoughtful and intelligent and, as a guest lecturer, I was heartened and satisfied with the outcomes and confident that the students with which I shared information would continue on their journeys of personal and political exploration and understanding.

The extent to which my presence in a classroom as a student affects others is often much different than my presence as a guest lecturer or panelist. I am now in my thirteenth month of transition and the physical changes, voice, body fat distribution, and facial hair are far more obvious than when I first began writing this article (summer 2004). During the 2004 spring semester at my university, I was still identified by peers and instructors as 'she'; particularly telling was a feminine voice and lack of facial hair. During that period, it was much easier to "come out" as transgendered, particularly in social justice courses (women's studies). The individuals in these classrooms are often prepared for diversity and for having their ideas of gender norms and corresponding social roles challenged. However, by the time I returned to classes in the fall of 2004, my physical identity, as well as my name, had/were changed so that I now easily present as 'he.' It has become more challenging and more involved to announce my gender identity and I have not come out publicly in any of my classes for the 2004-2005 academic year. Nevertheless, I have still participated in campus diversity forums, as well as delivering student presentations on gender identity. I have informed professors of women's studies classes in which I am enrolled of my trans identity. However, after a particularly unpleasant, anonymous phone message regarding my transness (received shortly after I participated as a student on a panel in a class on gender theory), I have been more cautious about disclosure. Although this adds a closet-dimension to my experience, I believe it still epistemologically provides me with a voice of authority. I now speak overtly as a male, but also covertly as a female. My insights into gender-related discourse often "surprise" both instructor and classmate–I believe being located at a pivotal point on the gender continuum gives me an opportunity to inform biological males of different ways to look at the world. However, "being heard" and listened to, while presenting as a male, may be nothing less than a by-product of male privilege–not a inconsequential variable.

As a lecturer or panel member, where I feel I can safely "come out" as an FTM, the totality of my life experience gives me an authority that few have at my university. On our campus, both my lesbian wife and I are the voice of trans-identity and all that goes with it. It is both a heady and frightening situation and one in which I must exercise constraint over both myself and the impressions heteronormative students/faculty members take away from our interaction. I feel a great responsibility to continually stress that my experience is not the experience of all transgendered people, regardless of the pressure from others to take on that role.

Over the span of my 55 years of life, my own journey has taken me many places, and in many different directions–from the only girl child of working class parents to the oldest of a very large and very broken family; from Indiana to Connecticut to California; from alcohol and drug addiction to nearly twelve years of living clean and sober; from a shadow-person with diminished self esteem, chronic depression and persistent self-destructive behaviors to a fairly well-centered, strong and determined, successful human; from a closeted teenage queer (in the pejorative form) to lesbian to butch dyke, to butch to transgender, to my present state as a female-bodied FTM queer. My experience in teaching and discussing transgender and related trans issues with those around me has been a huge part of my decision to carry through with transition. It was after my last transgender lecture given in the fall of 2003 that my partner and I, once again, began discussing the possibility of my transition. As stated above, this was a recurring theme for me, and the details of our decision to embark on this path could, as with anyone, fill many pages, so I will not discuss this process further in this forum. However, what is important to discuss are the many questions that this decision brings up. What do I want from my life? What are my academic goals? How will others see my new body? What do I call myself now? How can I maintain my queer identity while at the same time succeeding in projecting my inner core identity (not-woman, more male) to others? While presenting as a male with my high femme wife, how do I combat the world's natural assumption that we are heterosexual? How do I counter both internal and external discourses that proclaim I am no longer the person I was (not female socialized, not part of lesbian culture, not woman-identified)? How do I answer questions posed by future students, past lesbian friends and new others regarding my life choices and my presentation? What will be safe? What will be copping out? In what ways can I best serve my chosen queer community?

I have answers for some of these questions, though I know as an evolving human, until the day I die, things are always subject to change. I want to teach at the college level. I want to continue my research in the field of queer and transgender literature. I hope that those who see me will see me as I see myself–tall, handsome, sometimes wise, sensitive, aware and loving, married to a lesbian high femme as her queer feminist-husband. I believe that the subject of transgenderism and transsexualism is highly complex and deeply personal, and needs to be discussed intelligently and forthrightly among people of all ages, genders, classes, and races. I believe that sexuality, gender, and the present day dichotomous categories of female-male, woman-man, and feminine-masculine are self-limiting terms and poorly reflect our culture as a whole. At the same time I will defend the right of individuals who identify themselves with such terms, while continuing to work towards the goal of redefining the expectations, stereotypes and cultural imperatives placed on gender.

STUDENT OUTCOMES: WHAT DOES T MEAN TO ME?

Students usually begin the unit on transgender professing to know nothing about transgender life, to be completely confused about the different possible issues faced by transgender individuals, and to not even understand how it might be possible to actually cross gender lines (rather than merely to "perform" gender outside of expected norms). Many in my classes have expressed extreme discomfort with the idea of transgender and have stated that they would never consent to date a transgender individual. Their facial expressions, body language, and choice of words in the initial discussions demonstrate that they feel ill-at-ease, sometimes even angry about the subject, without truly knowing why. Over the course of the three-hour session on transgender identity, the students visibly relax. As we continue to bring up the issues related to transgender through the rest of the semester, they show a willingness to consider transgender as a viable set of options. They begin to look for evidence of how transgender relates to their own life and how they can work toward creating a society with a better understanding of genders.

With so much institutional focus on assessment, and given my own wish to incorporate student feedback in the constant development of my courses, I always plan some assignments or activities toward the end of the semester that allow the students to reflect upon course content and discuss how the course has changed them. When given the choice of which identity to define, many students have talked about transgender. This tells me that the transgender unit makes a strong impression on them, that the students feel that they have learned–and retained–something valuable, and that now the concept of transgender is comprehensible to them. Moreover, they show a remarkable sensitivity to the complexities of the transgender experience. From what the students have said, much of this is due to having the opportunity to "put a face" onto the transgender category, through accessible and personal readings, discussions with visiting speakers, and interactions with students who come out as transgender. Their comments on the final exam administered in the fall of 2004, some of which I will reproduce here, run the gamut from analytical to poignant.

- "An umbrella term, transgender gives identity to people who may be struggling to find just that."
- "Their identities are reduced to cartoons and they feel isolated from the community."
- "Changes in transgender politics are interrelated with changes in personal aspects of transgender identity. For example, if a person who is in a lesbian or gay relationship changes their [sic] gender or sex, the couple faces many changes in the way their relationship is perceived."
- "When looking at the personal challenges, we can see that it is also a matter of 'passing' in a hetero society. Oftentimes transgendered people must go from passing as either gay or straight to becoming another gender entirely. This can cause problems with jobs, family and partner status."
- "As Connor put it in class, classification under the DSM gives the transsexual a protocol for approaching their situation. While the requirements may be somewhat objectionable, they provide a framework for moving forward, they indicate a starting point in the process of transformation and they give the transsexual a verified, acceptable identity that cannot be argued by medical professionals."
- "How much is it really worth to be the sex you want to be, is it enough to make you broke and penniless?"

Students also have shown–in written assignments, class discussions, and in activist projects in the community–that they are personally affected by the issues covered in class. They are noticing a transformation in themselves and they are attempting to make a difference in how society views transgender. One student says, "I found it really great to learn about struggles within the Transgendered, specifically transsexual, communities. . . . And having AM [student alias] and Connor talk to us and express their experiences was important. . . . I have become more aware and proud of my own experiences and encouraged by the strength of others." Another student relates, "All the things I have learned in this class have better enabled me to back up my statements and help those in my own community be more accepting and open-minded toward those different than themselves." One straight-identified student said, "I feel like I need to take it upon myself to use my privilege against the injustice experienced by LGBTQ identified people."

This academic year I have seen some profound and moving incidents. One of the students came out as a transsexual and utilized the classroom space to validate herself and to educate her peers. She explains.

> The whole course was helpful to me, it made easier for me to come out to other students and made me safer and to know that such a course exists in a university is very relaxing to me. . . . I am a male-to-female transsexual and that part of the course really helped me get out what I needed to say to some people about my sufferings and the problems that I went through coming from a very conservative closed minded, third world country.[12] I contributed to the course by giving more information about transgendered people in the Arab world. (final exam)

Her courage and wealth of information was mentioned by many of her classmates; one young man asserts on his final exam that "Possibly the single most educational experience of this semester has been my personal experience speaking with our classmate AM." Inspired by the learning experiences of this course, students are taking up the issue of transgender in outside projects. One student who took the course last semester, who currently is serving as Assistant Director to the Associated Students Women's Center, is making the topic of transgender central to this year's events around Breaking the Silence. A student from two years ago, Julie Cosenza, directed Chico's production of *The Vagina Monologues* this semester and casted a MTF lesbian from the community to perform the monologue "They Beat the Girl Out of the Boy . . . Or So They Tried." This is the first time that a transgender individual was invited to perform in the Chico production, and the first time this monologue was included. The audience was visibly moved by the performance, and the actor reports feeling empowered by her participation; she already has agreed to be a key member of the organizing committee for next year's show. Cosenza affirms that this course, as well as a course on Theoretical Perspectives on Gender, in which she read *Transgender Warrior* by Feinberg, directly influenced her decisions in directing *The Vagina Monologues*.

THE TRANSFORMATIVE POWER OF TEACHING TRANSGENDER

Like my students and my "queer-feminist husband," I believe in the transformative power of teaching transgender—without a doubt learning about the transgender continuum can spark profound illumination and development for anyone. Connor has already addressed how teaching transgender made an impact on his decision to transition. For me, the process of teaching transgender along with my significant other has brought about life changes that I never would have imagined, which I will discuss here. In the first year of teaching this course, we identified firmly as a butch/femme couple who had seen several individuals in our community transition from female to male, with all of the problems and challenges that entails. We counted numerous tomboys, sissies, drag queens, and other gender deviants in the larger circle of our acquaintance. My partner spoke honestly in class about her longtime questioning of her gender assignment, and her eventual decision to remain a butch lesbian who enjoyed primarily the company of women. Through intensive research and lengthy conversations, we all came to a much deeper level of respect for and understanding of transgender individuals.

Over the next year, CJ and I began planning our wedding and became intimate friends with several community members who lived outside of the gender norm. One lesbian couple had originally met and married as man and woman, previous to the husband's decision to transition. Another friend is legally female, but does not identify as either woman or man, although she presents as more masculine than feminine; her/his partner is a high femme. That year when appearing as a guest lecturer in the class, my partner identified as transgender, in that s/he did not feel s/he truly fit into the category of woman or lesbian, but rather lived in transgression of gender norms. For me, this new gender designation was both interesting and exciting, if a bit scary; in truth, the different identification did not seem to change the way that we lived our lives.

Two years ago, my partner agreed to give a series of lectures in the LGBTQ course in order to receive senior-level credit in Women's Studies, and thus probed deeper than ever into the extant research on the subjects of Queer Comix, Russian Queer, and Transgender. As in years past, my partner's lecturing about transgender, along with our conversations before and after, brought up the regret that s/he had not been born ten or twenty years later, so that gender transition would have been a more viable option. In the classroom, as we both shared stories about transgender people we know and yet again spoke about how their lifelong dreams and self-perception had been vindicated, we started talking about it more at home. Several students that year (and, truth be known, in previous years) asked outright whether CJ had ever considered transitioning now. I think that the open reception by the younger generation and continual talks about the process of transition challenged her to face the fears and internalized transphobia that had been holding her back. So, when yet again I gave my assurances that I would still support a transition if that truly was her/his wish, s/he made the decision to try. That is when our learning process really began, and a couple of months later CJ began testosterone injections and resumed psychotherapy with a transgender specialist. To describe the changes that have taken place over the last year and a half or so would require at least another article, if not an entire book, but suffice it to say that our life has taken a radical turn. With this new development, the teaching of transgender has become increasingly relevant and meaningful for me. At the same time, living in this ambiguous space is more confusing and fraught with emotion than I could have envisioned; I sense a new hesitation and vulnerability in myself that make me wonder how I will approach the subject in the semesters to come. I know that this is an opportunity to eschew the superficialities and linguistic impoverishment of the sexuality/gender discussion lamented by Biddy Martin. Martin inspires me to bring into the classroom a "reconfiguring [of] the institutional and discursive conditions that structure and are structured by regulatory norms, but also reconfiguring interiorities, and, in particular, distributions of power, autonomy, attachment and vulnerability" (106). We are now in an ideal position to demonstrate *bodily* and *intellectually* how femininity and masculinity are not synonymous with male and female, how queer sexuality is (for us) less about who may have facial hair and much more about how one's life experiences intersect with one's corporeal existence. In the classroom we are able to explore the new tensions around visibility, discuss moments of panic or of abashed passing, and formulate strategies for education and activism along with the students.

I think that Connor is able to express the change much more eloquently: *My body has become a classroom, a teaching tool, a chalk-
board, a power-point presentation for any who wish to attend and to
learn from. My relationship has become a template upon which new
designs are being etched, from which new patterns for communication and loving are being drawn. My politics have become a testing
ground for feminist theory, gender role resistance, and intellectual
challenge, forcing me, my fellow students, my pedagogical colleagues to struggle with imperatives and ideas previously thought
written in stone.*

Having been so thoroughly inspired by my husband (to delve ever
more deeply into the research, to update and augment rather than
staying complacent in my knowledge and perspective) and students (to believe in the possibility of social change), I have continued to integrate Transgender into my own work. In 2003 I
presented a workshop on transgender in our CSUC Women's Conference, attended by people of all walks of life. At one point we
broke into groups in order to complete Bornstein's 10 Minute Gender Outlaw Exercise,[13] a task that provoked some resistance (happily,
most participants rose to the challenge) and some disturbed protest (one
woman felt that the lighthearted exercise minimized the serious nature
of transgender oppression). As a penultimate activity, we reviewed
the International Bill of Gender Rights, which was created by the
International Conference on Transgender Law and Employment
Policy at their second annual conference, held in Houston, Texas,
August 26-29, 1993 (and adopted June 17, 1995).[14] To end the workshop, we collectively envisioned ways that we could contribute to
the fight, making the political personal once again. Afterward, several people spoke of the transformative power of the conference in
general, and the transgender workshop in particular. For many, a
general sense of discomfort, the unease that stems from ignorance,
had been dispelled. For the first time, they felt they were part of a
larger struggle that included folks that before had seemed drastically
different, strangely fearsome, and threatening to their sense of reality. Education truly is the gift that keeps on giving.

As the course and I are changing, so is the institutional context for
teaching transgender at Chico State. As of May 2004, the course was
conferred General Education status and now can fulfill the requirement
for Ethnic Diversity and/or the Gender Perspectives Senior Theme.
Several faculty members have cooperated in writing a proposal for a
new LGBTQ Studies minor, which if approved could be the third

LGBTQ program in the California State University system, after the San Francisco and Humboldt campuses. We have found substantial support for this project by the faculty and staff coalition formed at the first annual CSU Queer Studies conference, held in 2004 at the Humboldt Campus, and where teaching transgender was one of the workshop topics.[15] As was discussed at length at the conference, we will need to continue to demonstrate student interest and prove the academic and social relevance of LGBTQ, in order that these programs not be sacrificed to the state budgetary crisis. For the nonce, the Women's Studies program has been shuffled over to the supervision of International Studies, so one does wonder how LGBTQ, much less the T itself, will fare when placed under yet another level of administration. But wherever there is a snarl, there are always new strategies to unravel it, and thus continues my particular saga of teaching transgender.

ISSUES AND QUESTIONS FOR FURTHER CONSIDERATION

1. What would be the elements of an effective textbook or anthology for teaching Trans? In a Queer Studies course? In a Lesbian Studies course? In a Social Justice course? In a general education course?
2. How can teachers of Trans material learn from the history of teaching LGB Studies? What are the direct parallels between the subject matters? What elements differ completely?
3. To what degree is an "interested positionality" (i.e., being or being in relationship with a Trans individual) perceived by students as necessary to infer authority on the subject?
4. How helpful are narratives, whether biographical, autobiographical, or fictional, in putting a face on the issue? Is this a key factor in engaging student interest?
5. How much teaching Trans should focus on individual narratives versus systemic issues, problems, and solutions?

AUTHOR NOTE

This article is presented as a union of two perspectives, of two voices. In the literary tradition of visibly marking multiple narrators, Connor Trebra's voice is presented in italics.

NOTES

1. To speak of a Transgender Movement is as slippery as to refer to Feminism in this day and age, in that both are umbrella terms that refer to broad sociopolitical ideologies rather than a specific set of objectives pursued by a defined group of individuals. In general terms, the Transgender movement may include the activism of scholar/activists like Les Feinberg and Kate Bornstein; the proliferation of organizations that support transgender individuals and educate the public (e.g., The National Center for Transgender Equality); the growing number of research groups and lobbyists that focus on positive social change around gender (e.g., the Transgender Law Project); the budding international cycle of conventions highlighting transgender issues (e.g., True-Spirit Conference in Laurel, MD); and the burgeoning Internet resources on the topic (see, for example, the DMOZ list of transgender-centered links: http://dmoz.org/Society/Transgendered/Activism/).

2. David Román cautions us that we need to take into account our own subject positions when teaching in gay and lesbian studies (113-14).

3. I am indebted to Kelly Coogan for helping me to flesh out this concept. For further elaboration, see her article *Fleshy Specificity: (Re)considering Transsexual Subjects in Lesbian Communities* (this volume).

4. Cherrie Moraga and Amber Hollibaugh remember that "lesbianism . . . came to be seen as the practice of feminism. It set up a 'perfect' vision of egalitarian sexuality, where we could magically leap over heterosexist into mutually orgasmic, struggle-free, trouble-free sex" (244). Nevertheless, the vision of utopia would create as many clashes as it solved. Judith Roof complains that in the 1970s "political zeal had locked its adherents into a species of puritanical sexual policing" (27). She explains that "since all things that were masculine were evil and all things that were feminine were good, sex between a man and a woman could only enact an oppressive situation, while sex between two women necessarily tapped into such positive womanist traits as gentleness, warmth, kindness, equality, decenteredness, non-goal orientation, and a total absence of dominant/submissive formations (which would include any relation of masculine and feminine, sadomasochism, or sexual activities such as tribadism or penetration that seemed to imitate heterosexual practice) were perceived as reproducing the oppressive relations of patriarchy" (28). Nor were these ideals "optional" in the women's community, as Lillian Faderman's history of lesbian life documents. "Despite the movement's rhetoric about love for all women, those who, by some infraction of the code, were judged 'politically incorrect' were given cold treatment by the community. Being politically correct (p.c.) meant that one adhered to the various dogmas regarding dress; money; sexual behavior; language usage; class, food, and ecology consciousness; political activity; and so forth" (230). In order to be accepted/acceptable as a lesbian, one's sex "must be consistent with the best of lesbian ethics" (250). The sex radicals of the 1980s and after would seriously challenge this policing, leading to the "sex wars" chronicled by several scholars, including Gale Rubin. Sex-positive activists characterized p.c. sex as "obsessively concerned with not 'objectifying' women and with promoting humdrum 'equal time' touching and cunnilingus; they found absurd the 'politically correct' notion that any kind of penetration was heterosexist" (Faderman 253).

5. Mary Kay Thompson Tetreault (1995) developed a measuring tool to discuss how far universities, publications, etc., have come in integrating Women's Studies ideas. Her 5 phases of inclusion have been used widely to evaluate textbooks, teacher

education curriculum, and faculty development seminars. Kim Chuppa-Cornell explains the phases as follows:

Phase (1) Male Scholarship: represents the male experience as the universal.

Phase (2) Compensatory Scholarship: notes women's absence and adds a few outstanding women, which often reinforces the notion that most women are inferior or less developed intellectually outside a few exceptions.

Phase (3) Bifocal Scholarship: explores human experience in gender dualities, emphasizing differences between men and women as groups. Due to this emphasis on gender differences, material in Phase 3 runs the risk of devaluing women's separate sphere, overemphasizing women's oppression and passivity, and/or overlooking significant differences among women.

Phase (4) Feminist Scholarship: puts everyday women and their lives in the center of study. It asks new questions regarding how race, ethnicity, social class, marital status, and sexual orientation shape women's experiences. Scholars in this phase draw on multiple disciplines in order to understand such multifaceted experiences that often defy traditional periodizations and categorizations.

Phase (5) Multifocal, Relational Scholarship: continues the work of Phase 4 but also combines questions regarding men and women and the relationships between their experiences. This pluralistic and holistic phase includes multiple variables for analysis and looks for continua of humanness in men's and women's experiences versus the separate spheres approach found in Phase 3 (24).

Tetreault's Phases seem to me to be perfectly adaptable to an evaluation of how far LGBTQ Studies have come in universities. In terms of LGBTQ Studies I see us to be between stages 2-3 at Chico State (with some faculty members being far advanced, of course), attempting with the Sexual Diversities Minor to rise up to Phase 4.

6. Two other colleagues, Elizabeth Renfro and Terri Thomas Elliot, have been instrumental in changing the institutional lens through which gender is perceived and taught at Chico State. English professor Carol Burr, for many years the Director of Multicultural and Gender Studies (MCGS), also has been exceedingly forward thinking and always is willing to engage in enlightening conversations on diversity of all sorts. A number of faculty whose specialty lies outside of transgender have nonetheless helped carry the discussion forward, especially Jerry Maneker and Liahna Gordon.

7. Kim Chuppa-Cornell speaks at length about the challenges that emerge as scholars and teachers move from one phase to the next. According to her classification, this essay would be the articulation of a combination of "phase conflicts in which the author describes the experience of operating at a phase different from his/her students, materials, and/or colleagues" and "phase application issues in which the author discusses his/her experiences and obstacles teaching from a particular phase" (28). One obstacle that Chuppa-Cornell's research has focused upon and which describes my own experiences in the classroom is that "even if we [teachers] can overcome certain kinds of differences (race, gender) by focusing on certain kinds of similarities (social, political) among our students . . . those differences may still intervene . . . and once we make them conscious, suspicions and resistances may be aroused" (Caughie 1992, 784 qtd. In Chuppa-Cornell 31).

8. Barry Adam's *The Rise of a Gay and Lesbian Movement* has been an excellent historical resource for these lectures, as have Paula Rust's "Sexual Identity and Bisexual Identities" and Amanda Udis-Kessler's "Identity/Politics: Historical Sources of the Bisexual Movement" in the Beemyn/Eliason *Queer Studies* anthology.

9. Leslie Feinberg made famous the term Transgender Warriors, explained in several works, including: *Transgender Liberation* and *Transgender Warriors*.

10. See Kate Bornstein's book *Gender Outlaw: On Men, Women, and the Rest of Us* (New York: Vintage, 1994).

11. Paulo Freire defines dialogics, or the integration of both action and reflection in dealings among "leaders" and "the people," as an instrument of liberation by which individuals and groups are able to fully experience their humanness (Chapter 4: 106-164).

12. In the classroom I attempt to create a space free of vitriolic, personal attacks, or unsubstantiated derogative claims. Also, I want students to develop their critical acumen and skills of debate, so I challenge them to explain and defend even positive remarks or observations about LGBTQ. At the same time, I wish to acknowledge the reality of the dangers inherent in being queer and living in a rural agricultural community. In this, my methods reflect those discussed by Harriet Malinowitz in *Textual Orientations* (112-13).

13. Participants were encouraged to "answer one or more of the following questions, writing down a series of answers: What is a Man? What is a Woman? Why do I have to be one or the other? The trick is that the answers have to be phrased in questions." I gave fragments of the examples provided by Bornstein: "What is a Man? What's a Woman, for that matter? What's a boy? Was I ever a boy? What was it like to be treated like a boy? Did I like it? What did I like about it? How do I like to be treated today? Does that make me a boy, still?" Page 32 *My Gender Workbook*.

14. The short version of the International Bill of Gender Rights follows. Unfortunately, the official ICTLEP Website is not accessible at the time of this writing. However, the full text of the bill, where one may easily see the inspiration of other human rights movements, is available online at: *http://www.altsex.org/transgender/ibgr.html* International Bill of Gender Rights (as adopted June 17, 1995 in Houston, Texas, U.S.A.).

1. The Right To Define Gender Identity
2. The Right To Free Expression Of Gender Identity
3. The Right To Secure And Retain Employment And To Receive Just Compensation
4. The Right Of Access To Gendered Space And Participation In Gendered Activity
5. The Right To Control And Change One's Own Body
6. The Right To Competent Medical And Professional Care
7. The Right To Freedom From Psychiatric Diagnosis Or Treatment
8. The Right To Sexual Expression
9. The Right To Form Committed, Loving Relationships And Enter Into Marital Contracts
10. The Right To Conceive, Bear, Or Adopt Children; The Right To Nurture And Have Custody Of Children And To Exercise Parental Capacity

15. The conference organizers, Eric Rofes and Nan Alamilla Boyd, with the help of conference participants, have facilitated the dissemination of syllabi, program proposals, bibliographies, and other useful documents.

REFERENCES

Adam BD (1987). *The Rise of a Gay and Lesbian Movement*. Boston: Twayne.

Beemyn B and Eliason M, Eds. (1996). *Queer Studies*. New York: New York University Press.

Bornstein K (1994). *Gender Outlaw: On Men, Women, and the Rest of Us*. New York: Vintage.

Bornstein K (1998). *My Gender Workbook: How to become a real man, a real woman, the real you, or something else entirely*. New York: Routledge.

Brant B (1985). *Mohawk Trail*. Toronto: The Women's Press.

Brown ML and Rounsley CA (1997). *True Selves: Understanding Transexualism (For Families, Friends, Coworkers, and Helping Professionals)*. Philadelphia: Temple University Press.

Caughie PL (1992). Not entirely strange. . . . Not entirely friendly: passing and pedagogy. *College English 54*(7), 775-93.

Chuppa-Cornell K (2005). The conditions of difficulty and struggle: A discovered theme of curriculum transformation and women's studies discourse. *NWSA Journal 17*(1), 23-44.

Cromwell J (1999). *Transmen & FTMs: Identities, Bodies, Genders & Sexualities*. Chicago: University of Illinois Press.

Devor H (1997). *FTM: Female-To-Male Transexuals in Society*. Bloomington: Indiana University Press.

Faderman L (1991). *Odd Girls and Twilight Lovers: A History of Lesbian Life in Twentieth-Century America*. New York: Columbia University Press.

Feinberg L (1992). *Transgender Liberation: A Movement Whose Time Has Come*. New York: New World Forum.

Feinberg L (1996). *Transgender Warriors: Making History from Joan of Arc to Ru Paul*. Boston: Beacon.

Freire P (1993). *Pedagogy of the Oppressed*. Trans. Myra Bergman Ramos. New York: Continuum.

Grace AP and Hill RJ (2004). Positioning queer in adult education: Intervening in politics and praxis in North America. *Studies in the Education of Adults 36*(2), 167-189.

Halberstam J (1998). *Female Masculinity*. Durham, NC: Duke University Press.

International Conference on Transgender Law and Employment Policy. *International Bill of Gender Rights*. AltSex: Alternative Sexuality Resources. http://www.altsex.org/transgender/ibgr.html.

Israel GE and Tarver DE (1996). *Transgender Care: Recommended Guidelines, Practical Information & Personal Accounts*. San Francisco: Jossey Bass Books.

Malinowitz H (1995). *Textual Orientations: Lesbian and Gay Students and the Making of Discourse Communities*. Portsmouth, NH: Heinemann.

Martin B (1994). Sexualities without Genders and Other Queer Utopias. *Diacritics: Special Issue on Critical Crossings 24*(2-3), 104-21.

Moraga C and Hollibaugh A (1992). What we're rollin' around in bed with: sexual silences in feminism . . . a conversation toward ending them. In: Nestle J (Ed), *The Femme-Butch Reader*. Boston: Alyson, pgs. 243-53.

Namaste VK (1996). Tragic Misreadings: Queer Theory's Erasure of Transgender Subjectivity. In: Beemyn B and Beemyn ME (Eds.), *Queer Studies: A Lesbian, Gay and Bisexual Anthology*. New York: New York University Press, pgs. 183-203.

Nanda S (2000). *Gender Diversity: Cross-Cultural Variations.* Prospect Heights, IL: Waveland.

Román D (1995). Teaching differences: theory and practice in a lesbian and gay studies seminar. In: Haggerty GE and Zimmerman B (Eds). *Professions of Desire.* New York: Modern Language Association of America, pgs. 113-123.

Roof J (1998). 1970s lesbian feminism meets 1990s butch-femme. In: Munt SR (Ed.), *Butch-Femme: Inside Lesbian Gender.* London: Cassell, pgs. 27-35.

Rubin H (2003). *Self-Made Men Identity and Embodiment among Transsexual Men.* Nashville, TN: Vanderbilt University Press.

Taste This (Performance Group), Coyote IE, Eakle Z, Montgomerey L, Camilleri A (1998). *Boys Like Her: Transfictions.* Vancouver: Press Gang Books.

Wilchins RA (1997). *Read My Lips: Sexual Subversion and the End of Gender.* Milford, CT: Firebrand.

Developing an Identity Model
for Transgender and Intersex Inclusion
in Lesbian Communities

Christopher Robinson

SUMMARY. The article briefly highlights past ideologies of Essentialism and Constructionist thought, identifying their weaknesses as arguments of exclusion. By combining corporeality and mentality, the article posits an identity model, the spherical characterization model that argues for transgender inclusion in lesbian space by deconstructing the ideology behind identity group construction. Next the article applies the spherical characterization model to the characters in Virginia Woolf's *Orlando* and Jeffrey Eugenides' *Middlesex* to delineate how the model argues for transgender inclusion within lesbian space. *[Article copies available for a fee from The Haworth Document Delivery Service: 1-800-HAWORTH. E-mail address: <docdelivery@haworthpress.com> Website: <http://www.HaworthPress.com> © 2006 by The Haworth Press, Inc. All rights reserved.]*

Christopher Robinson has a bachelor of science degree in Science, Technology and Culture from The Georgia Institute of Technology. He plans to pursue doctoral studies in Sociology, specializing in Gender Theory and Social Corporeality.

Address correspondence to: Christopher Robinson, 419 Sinclair Avenue, Atlanta, GA 30307 (E-mail: crobinson542@hotmail.com).

[Haworth co-indexing entry note]: "Developing an Identity Model for Transgender and Intersex Inclusion in Lesbian Communities." Robinson, Christopher. Co-published simultaneously in *Journal of Lesbian Studies* (Harrington Park Press, an imprint of The Haworth Press, Inc.) Vol. 10, No. 1/2, 2006, pp. 181-199; and: *Challenging Lesbian Norms: Intersex, Transgender, Intersectional, and Queer Perspectives* (ed: Angela Pattatucci Aragón) Harrington Park Press, an imprint of The Haworth Press, Inc., 2006, pp. 181-199. Single or multiple copies of this article are available for a fee from The Haworth Document Delivery Service [1-800-HAWORTH, 9:00 a.m. - 5:00 p.m. (EST). E-mail address: docdelivery@haworthpress.com].

Available online at http://www.haworthpress.com/web/JLS
© 2006 by The Haworth Press, Inc. All rights reserved.
doi:10.1300/J155v10n01_09

KEYWORDS. Identity modeling, lesbian identity, lesbian separatism, intersex, literary criticism, gender theory, sexual variation, corporeal philosophy, Virginia Woolf, Jeffrey Eugenides

To actualize a lesbian community requires a finite definition for the lesbian body, the body being quantified in terms of being both a passive physical form and a tool of social interaction. The problem with using a static body in creating a definitive lesbian community is how the broad idea of lesbianism can be focused into the distinct physical and cultural definition of being a lesbian. The answer proposed by lesbian feminists in the past centered on an essentialist concept of identity, notably summed up under the term *womyn-born womyn* as a fundamental quantifier for the lesbian identity (in other words, only people born under the biological category of being a girl are capable of becoming a lesbian). Such terminology is inherently exclusionary because it focuses on creating a community centered on a static body encoded with privileged cultural experiences. The problem with this essentialist practice is that it fails to take into account the constructionist facet of social interaction. Such essentialism instead supports mind/body dualism by denying the performative gendered body and promoting a normative dichotic structure of sex.

In order to remove exclusionary practices of lesbian feminist communities, the essentialist definition of lesbianism focused on a privileged body ought to be replaced by one that allows for the inclusion of transgender and intersex individuals. But, simply shifting the definition of a lesbian body to a constructionist argument (i.e., arguing lesbianism is a matter of personal choice) presents the same problems of discourse as the essentialist definition in continuing to support mind/body dualism by linking experience with the body where the link between experience and the body comes from a continual privileging of certain bodies. However, any solution that develops a specific lesbian body for the inclusion of transgender and intersex individuals creates a quantifiable definition of lesbianism. The problem with having a quantifiable definition of lesbianism is that it automatically creates an exclusionary community in that the constituted subject, the lesbian, relies on an abjected[1] and outside subject for self-definition. In this way, simply redefining the lesbian body to include transgender and intersex individuals does not eliminate the problem of exclusion: it only reapportions it.

Overcoming the exclusionary rhetoric of a lesbian feminist community requires the erasure of mind/body dualism. A preferred approach considers lesbian identity a process of both essentialist and construc-tionist elements acting on the individual to shape identity uniquely. Viewing the central element of personal identity and social interac-tion, I posit that both essentialism and constructionism play roles in aligning the individual with a group identity. Furthermore, lesbian communities should include transgender and intersex individuals be-cause an exclusionary definition fails to recognize the specificity of identity. To represent my concept of identity, I propose the spherical characterization model, which illustrates a combination of essentialist and constructionist thought while highlighting individual identity as a coalescence of mind and body. Analyzing the fictional characters within the novels *Orlando* and *Middlesex*, I demonstrate the application of the spherical characterization model to specific transgender and intersex individuals. Applying this model to the characters supports their inclusion in lesbianism, not because they fit within a specific defi-nition of membership, but because they challenge the notion that quanti-fiable identities exist. This challenge occurs because both inclusion and exclusion can be demonstrated under essentialist and constructionist ideologies; however, identity inclusion within these ideologies results in exclusion towards other identities or creates hierarchies within the identity, but viewing the characters through the discourse of my model allows for inclusion without exclusion.

ANDROCENTRISM:
PROBLEMS WITH AN ESSENTIALIST BODY

Central in the field of gender and identity studies is the role of the body in terms of the effect its appearance has on social interaction and how its interior biology shapes behavior and desire. Essentialist lesbian feminism relies on biology shaping identity using the rhetoric of an in-ner essence as identity to exclude transgender and intersex individuals from events such as the Michigan Womyn's Music Festival.[2] Even though the *womyn-born womyn* policy of the organizers was made un-der the good intention of creating a protective and productive women-only space, using an essentialist argument for exclusion is an appropria-tion of a false patriarchal technique designed to create social hierar-chies under the purview of *natural order*. Sandra Bem's term androcentrism, indicating that social experiences and cultural artifacts are constructed through the male-gaze and how male experience is privileged over female (1993: 41), is a good example of an essentialist patriarchal structure that illustrates the larger social impact a naturalized exclusionary practice can have.

An example of androcentric practice is the devaluation of the female mind, stemming from mind/body dualism that separates men as the mind of society and women as the body. Since the ancient Greeks, the product of the mind, thought, has been privileged above the brute labor of the body. According to the tenets of mind/body dualism, women are connected to the body because of their role in heterosexual reproduction and child rearing, leaving men to be associated with the mind. The androcentrism of the situation arises because the mind-male connection and body-female connection have been *naturalized* through centuries of acculturation resulting in the philosophical hierarchy of mind/body dualism *naturally* valuing men above women. The general social significance of this discourse is seen through the lack of respect, recognition, and attainment female intellectuals received in the past, as well as the continual skepticism of female professionals in some circles.

The Michigan Womyn's Music Festival's policy of *womyn-born womyn* creates the same essentialist hierarchy in lesbian communities as androcentrism does in a patriarchal society. Privileging *womyn-born womyn* as the definition of woman not only excludes transgender and intersex individuals from participation but dehumanizes them by denying them the social label of their identity.

THE ONE-SEX MODEL:
PROBLEMS WITH A CONSTRUCTIONIST BODY

A clear example of a constructionist viewpoint on the body is the one-sex model. Originating with the ancient Greeks, the one-sex model's basic tenet is that the female body is an inversion of the fully formed male body and therefore less perfect. This belief, with small variations, remained the dominant understanding of human biology until the seventeenth century (Lacquer 1990: 25). Defining the female body in terms of the male body allowed men to justify their social dominance over women as *natural*.

The one-sex model is constructionist because it was primarily a tool for social sorting. In other words, the biological distinction of sex is not a central issue because the model does not directly seek to validate the male body over the female body; rather the model represents the male body as the human standard and the female body as an imperfect variation. Classification of sex within the model serves primarily to uphold the social hierarchy of gender according to the culture outside of the body. Thomas Laqueur illustrates the constructionist nature of the model in his discussion of legal cases involving the sexual classification of intersexed individuals. Laqueur writes, "The concern of magistrates was less with corporeal reality–with what we would call sex–than with maintaining clear social boundaries, maintaining the categories of gender" (1990: 135).

The one-sex model demonstrates the problems that would arise if intersex and transgender individuals were included in lesbian communities under a purely constructionist definition of lesbianism. An inclusive constructionist ideology of lesbianism would instead dismiss the body's importance for identity construction in favor of defining the lesbian body according to its subscription with a set of social qualifications. A constructionist model also creates a hierarchical system within lesbianism because individuals could feel required to participate in certain activities in order to obtain the social label of lesbian. Constructionism runs the risk of creating a process of initiation that would serve to further support a hierarchy in lesbian communities as individuals could only validate their identity through adherence to arbitrary constraints.

THE CORPOREAL BODY AND MENTAL SELF: COMBINING ESSENTIALIST AND CONSTRUCTIONIST IDEOLOGY

Transgender and intersex individuals present the ultimate challenge to the essentialist/constructionist dichotomy. Transgender people struggle, as the social meanings and interactions expected of their bodies contradict the internal desires of their identities. Intersex people find their bodies fall outside of the corporeal binary and subsequently suffer identity crises as they try to fit themselves within a dichotomous culture. Both essentialist and constructionist ideologies can place transgender and intersex individuals within lesbian communities, but inclusion solely under either ideology risks creating an arbitrary hierarchy within lesbian communities that is as damaging to the lesbian identity as the patriarchal structures outside of lesbianism. As ideologies, essentialism and constructionism have the power to create static solutions to problems; however, lesbianism is not a static problem but a dynamic social discourse. Hence, inclusion based on a static ideology will only ever be a temporary fix since the isolated problem that is solved is not actually a problem but the increase of human diversity calling for new sets of discourse.

The hierarchies arise within essentialist arguments because identity is emphasized as an intangible product of the mind that is reliant on innate biological elements. Conversely, hierarchies in constructionist models exist through a heavy reliance on social meanings applied to malleable corporeal interaction and appearance. In order for lesbian communities to avoid inclusive practices that simply reshuffle hierarchies around a new fundamental corporeal or social difference, the corporeal body and mental self have to be merged under an identity discourse that combines essentialism and constructionism. Going further to erase mind/body dualism by redefining identity as the merged or singular structure of the corporeal body and mental self, the spherical characterization model presented here can be realized as a model of identity that combines essentialism and constructionism; whereby the combination is not simply a mashing together of ideologies but a utilization of essentialism's innateness and constructionism's sociality to engender a new discourse for identity.

AN INCLUSIVE MODEL OF IDENTITY

When conducting my research, the main conceptual problem I had with binary models was their linear structure. It seemed the expansion of binary models to include gradations between polar ideals resulted in poor representation for an actual social situation. The Kinsey model seemed to be the best model for progressive variation between two points, but a simple range of percentages ignores the possibility of variation within any certain event. More specifically, simply stating a person has a 10% chance of being gay or is 37% more likely to engage in a homosexual encounter ignores not only the individuality of the person but the uniqueness of each situation that the person encounters.

My first concept of expressing identity as situationally explicit rather than a general prediction of choices was to have a circle of behaviors where the person acting on them was placed at the center point. My reasoning behind this model was looking at specific actions, i.e., kissing a guy, as the same social occurrence not quantified by the actors involved. If the action of kissing a guy is static, then the interpretation of meaning changes based on the identity of actor, i.e., is the guy being kissed by a man or a woman is what gives the event meaning because it categorizes the identity of the actor (one who kisses the guy) as homosexual or heterosexual.

The problem I encountered with simply taking a linear concept of behaviors and pushing them together to form a circle was that the corporeal body was required to be a static object in the center in order for actors' identity to have meaning; this implies a certain innateness of the body too closely tied to essentialism and becomes problematic because it still allows for the privileging of certain bodies. In order to make the corporeality of the body dynamic, it was necessary to take the two-dimensional circle and turn it into a three-dimensional sphere, this way the corporeal body becomes a dynamic element expressible as specific points on the sphere while identity is expressed as a unique situational construction lying at the center of the sphere (I term this point the Body). The Body is the location of individuals that gives each identity an equal possibility of expressing any point on the sphere. Hence I arrived at the discourse termed the *Spherical Characterization Model*.

The Spherical Characterization Model[3] represents how the Body[4] uses both constructionism and essentialism in identity construction. The spherical characterization model places the Body at the center of a sphere on which every possible form of human behavior is represented by a point of the sphere's surface. Individual identity is constructed using rays to connect the Body to points on the sphere.

The Body within the Spherical Characterization Model is both the corporeal and intangible self because the two are not truly separable through cultural discourse. Furthermore, the mind and the corporeal body cannot be separated because the corporeal body is the element of interaction, or "the body is not passive in socialization" (Gatens 1996: 10). The Spherical Characterization Model views the corporeal body not as a limit of its physical dimensions but as the object by which the individual navigates through social space. Mind/body dualism is eliminated through the model because the constructionist nature of corporeal interaction prohibits the formation of identity outside of the body; in other words, a disembodied mind as being the actual self is no longer possible, the self is a coalescence of mind and body.

Conception of the Body within the Spherical Characterization Model is essentialist because the model holds that each individual is born with innate preferences to certain behaviors. At the same time, the Model's representation of the Body expresses that each individual is equally close to every behavior, albeit not equally likely to express it. Because each individual Body is different, the model resists viewing the Body as a *tabula rasa*, positing instead that there is a certain amount of innate processing for identity construction. This form of behavioral representation differs from sex-linked gender ideology[5] because it does not group the acquisition of behaviors based on the classification of the body; rather the model allows individuals to have behaviors that are mutually exclusive under the conventional gender dichotomy.

The constructionist elements of the model are the points of the sphere because they are defined by physical and cultural limitations. For example, there is no point for flight because, regardless of how much an individual feels they should be able to fly, the human body can't fly; however, this instance is not a finality because if biological science finds a way for humans to fly, the point would be added to the sphere.

Another important constructionist aspect of the Spherical Characterization Model is that socialization is needed for the cognition of an identity. Without society, an individual identity cannot develop because the Body by itself neither carries internal meaning, nor can it acquire meaning without social interaction. In other words, an external social structure is necessary to construct identity because the cultural context provides the tools necessary for self perception.

The Spherical Characterization Model uses the same sphere of behaviors for each individual, breaking the corporeal body away from dichotic ideology by removing all classification of gender to view the Body individually instead. Within the Spherical Characterization Model, gender identity is no longer based on the classification of gender schemas but eliminated entirely.

Sex as a congenital identification is eliminated because the biological functions that classify its distinctions are removed from the corporeal body and placed as points on the sphere. While this may appear paradoxical, the female body is still not able to act as the inseminator, it is necessary for social equality because gender schemas will continue to exist unless sex identification is understood to be important only as a facet of social hierarchy. Furthermore, the dichotomy of sex is broken down because conventional views of male and female are represented as being equally as natural as *5-alpha-reductase deficiency* (this condition is addressed later as it pertains to *Middlesex*) and other intersex variations.

Through the removal of corporeality from the Body, the Spherical Characterization Model eliminates the need for a *third sex* and helps support the Intersex Society of North America's policies against infant genital surgery (Turner 1999). Infant surgeries are performed primarily so that the child can grow up and acquire a *stable* gender identity within the conventional gender dichotomy,[6] but the Spherical Characterization Model represents how it is unnecessary for an intersex body to be surgically altered for the individual's identity to develop.

The removal of sex from the corporeal body also illustrates how the tangibility of the body does not automatically shape identity. This principle is demonstrated within conventional ideology by examining transsexual discourse. At first transsexual discourse seems to support sex-linked gender ideology because transsexual individuals have a desire to match physical sex to gender identity. But Nan Boyd notes that the substantiation of sex-linked gender is not exactly what is occurring because transsexuals tend to form a highly stereotypical gender identity to support their body morphic desire (Boyd 1997). Transsexual discourse as an argument against the conventional sex and gender-linked identity is further supported by the tactic utilized by twentieth-century transsexuals, where they commonly recited the transsexual narrative found in medical texts to qualify for surgical procedures. The fact that such measures were necessary to justify personal identity when confronting medical gatekeepers reveals the danger in creating communities around quantifiable elements.

The gender identity of transsexuals as falling within the conventional binary is speculative because the behaviors acquired for admittance into the community are not necessarily the product of an individual's innate desires but a visualization tactic used to justify the social perception of their physical alteration. Attempting to align personal conception to social perception results in highly stereotypical behaviors in some individuals because they think that in order for society to see them as a *real man* or a *real woman*, they have to exhibit the ideal behavior schemas of that sex identification. Conventional gender schema acquisition also leads to identity problems within transsexual communities. As Douglas Mason-Schrock discusses, transgender individuals can face an identity crisis towards inclusion in transgender communities if they enjoy activities that are highly associated with the sex they want to change from (1996). Most notably, Mason-Schrock mentions male-to-female transsexuals who feel conflicted at justifying their decision to become females as adults if they engaged in typically masculine activities as children; specifically, Mason-Schrock notes athletic ability or enjoyment in playing sports.

Whether transsexuals make their identity change by altering their appearance with stereotypical gender schemas or surgical operations, it is ultimately only the superficial elements of cultural classification that are altered. The Body of the individual does not change, because the erasure of mind/body duality within the spherical characterization model results in a plasticity of corporeality that allows individuals to undergo physical alterations as representational of their identity not as an alteration of identity as seen under a binary ideology. From the trans-sexual viewpoint, they are aligning the superficial visualization of their corporal bodies to the internalized concept of their *true self* (Mason-Schrock 1996).

In terms of exclusionary lesbian communities, the Spherical Characterization Model demonstrates that, even with shared experiences, individuals will have different structures of identification because they have different Bodies; therefore, the *womyn-born womyn* argument is invalidated as a quantifier because no two *womyn-born womyn* can have the same life experience.[7] Furthermore, the positioning of the Body within the center of the sphere equalizes the possibility of any experiential difference resulting in the behavioral point of lesbian being an equally valid identity for a *womyn-born womyn*, transgender, or intersex person. Transgender and intersex individuals reveal that identity does not confine itself nicely within quantifiable groups; rather as Kate Bornstein wrote about herself, identity is a *collage* (1995). The Spherical Characterization model provides the visualization of identity structure as being a collage as opposed to the clear division of sets that exist in dichotic linear ideologies. The Spherical Characterization Model also allows for the inclusion of non-op transsexuals and (re)constructed intersex individuals[8] within lesbianism.

HOW ORLANDO INTERNALIZES THE CHANGE IN HER BODY AND OTHERS' INTERACTIONS TOWARD HER

Orlando is an idealized transsexual text because the main character undergoes a spontaneous and complete sex change. The reader is informed of this change by a remark the narrator makes about Orlando's body. "He stood upright in complete nakedness before us, and while the trumpets pealed Truth! Truth! Truth! we have no choice left but confess–he was a woman" (Woolf 1956: 137). Because the novel predates genetic science, sexual identification is based only on appearance, specifically that of the genitalia. Viewing *Orlando* as a transgender text reveals the static nature of identity as being inseparable from corporeality. Orlando herself illustrates how physical changes do not alter an essential Body but do lead to a change in behavior and cultural interaction.

Orlando begins life as a boy: "He–for there can be no doubt of his sex . . ." (Woolf 1956: 13), and grows up assuming a masculine gender script thinking nothing about why he acquires certain traits until after his metamorphosis, when he is forced into a feminine gender schema.

Orlando is acutely aware that the alteration of his sex has done nothing to change his personality. Becoming a woman results in Orlando's gaining an understanding that the restrictions placed on women by men are made under faulty assumptions of what women are *naturally* like. Orlando's knowledge is akin to the realization of transsexual and intersex people that terms like *womyn-born womyn* do not encompass the reality of identity structures but serve rather to define a faulty ideal for group inclusion/exclusion.

One of the first things Orlando realizes after the sex change is the permanence of his character: "But in every other respect, Orlando remained precisely as he had been. The change of sex, though it altered the future, did nothing whatever to alter their identity" (Woolf 1956: 138). This affirmation on the static nature of the Body supports the essentialist concept that behavioral characteristics are innate. Orlando's having to learn feminine behaviors indicates the artificiality of the dichotic classification system. However, because corporeality plays a central part in sex-linked gender ideology, Orlando is forced to conform to a new gender schema since some of her old behaviors are inappropriate according to the social perception of her new body. The process of changing gender schemas allows Orlando to make observations on gender through the comprehension of someone who has learned and been accepted in both dichotic classifications. This learning process illustrates how society enforces strict dichotomy and penalizes nonconformity.

Applied to lesbian communities, Orlando's changing behaviors illustrate how constructionist ideology views identity as a social composite of specific behaviors and appearances. Furthermore, Orlando's forced acquisition of a feminine gender schema is similar to the ideology in excluding transgender individuals from lesbian feminist communities on the basis that a male-to-female individual will never be able to fully remove her innate maleness. The danger of conformity in such a way, as Orlando demonstrates, is that the individual feels it necessary to compromise parts of her identity in order to become part of the group. Communities that require individuals to compromise individual identity to preserve group participation, albeit implicitly, risk creating rifts; individuals are forced to continually validate their inclusion.

By looking at her past sex through the eyes of her present sex, Orlando becomes disoriented about her place in society. Because experience has allowed her to transcend[9] binary thinking, she lacks a standardized view of the world: "And here it would seem from the ambiguity in her terms that she was censuring both sexes equally, as if she belonged to neither" (Woolf 1956: 158). By criticizing humanity as a detached transcendental observer, within both cultural and temporal space, Orlando indicates the artificial nature behind the dichotomy of sex-linked gender ideology, demonstrating that people actually think outside of a polarized system but are constricted within them by acculturation. This concept is important for the inclusion of transgender and intersex individuals in lesbian communities because it illustrates how a quantifiable group definition inevitably confines social thinking into distinct classes by ignoring the infinite variations in human identity.

WHY ORLANDO FINDS IT NECESSARY
TO ALTER HER BEHAVIOR

Even though Orlando comprehends the arbitrary nature of sex and gender, she still adapts to the feminine gender script out of social necessity. Orlando's conscious process in deciding to alter her gender script signifies the constructionist nature of gender acquisition. Orlando initially finds herself with a transcendental consciousness of male and female when, after leaving the gypsies, the boat finally reaches London. On deck with the captain, Orlando is overcome by grief when she learns of a plague and fire in her absence. "Do what she would to restrain them, the tears came to her eyes, until, remembering that it is becoming in a woman to weep, she let them flow" (Woolf 1956: 165). Orlando visibly presents her emotions in this scene because is it expected of a woman and supports the cultural construction that women need the protection of men. But Orlando is reluctant to give up all of her masculine traits, desiring to retain the independence she had as a man: ". . . if it meant conventionality, meant slavery, meant deceit, meant denying her love, fettering her limbs, pursing her lips, and restraining her tongue, then she would turn around with the ship and set sail once more for the gypsies" (Woolf 1956: 163). This affirmation is made before she cries, and she does retain an independence of character throughout the novel, which is highlighted by her refusal to marry immediately upon returning to England.

Despite her earlier convictions to the contrary, Orlando eventually develops a need to marry. At first she simply buys a wedding band and sticks it on her finger to pretend. Then, she meets Sheldon; they fall in love; and marriage as a reality enters her mind. The romance between Orlando and Sheldon serves to finally signify to Orlando that she is a woman because in addition to adopting feminine behaviors and appearance, she is now finally engaging in a heterosexual relationship. "'I am a woman,' she thought, 'a real woman, at last'" (Woolf 1956: 253). Ironically, it is shortly after she and Sheldon become involved that her court cases are settled decreeing her a woman as well. The significance of these two events is that it is the answer of sexuality that finally resolves the question of sex identification. The fact that Orlando finally takes on a sexual role serves to validate her not only as a woman but as a potential mother; therefore, she becomes female. After Orlando officially becomes a woman and wife, she also becomes pregnant, and at the end of the novel, she is a London housewife running errands for her family, compulsorily aligned with idyllic heterosexuality.

Orlando demonstrates how sex identification of the body is used to impart identity restrictions on the individual in terms of adherence to cultural ideology. Within the Spherical Characterization Model, Orlando reveals a permanence of character that is maintained through her manipulation of cultural artifacts. Specifically Orlando uses dress to represent herself as male in order to enjoy the freedom of mobility privileged to men in her society. This permanence also reveals how corporeal alterations do not change the desires of the Body but do impact the social navigation of their fulfillment. *Orlando* argues for transgender inclusion in lesbian space because Orlando does not leave behind her *essential maleness*; rather, her identity becomes a fluid composite of her experiences as both sexes, illustrating that dichotic social systems do not represent the variation of identity. Orlando's reluctance to acquire a feminine gender schema demonstrates that quantifiable groups are based on restrictive social structures.

MIDDLESEX:
HOW CLASSIFICATION OF THE BODY DETERMINES CULTURAL UNDERSTANDING AND SOCIAL INTERACTION

Like *Orlando*, *Middlesex* spans multiple generations through the transcendental eyes of Cal: the narrator and main character. Cal, born Caliope, has the genetic variation *5-alpha-reductase deficiency* that results in him being born with externally appearing female genitalia despite his XY genetics. We begin the story with Cal as an adult recounting the events that lead up to his unique genetic condition starting with his grandparents. The novel spans the lives of two generations before Caliope is born. Jeffrey Eugenides, the author, makes the plot explicitly clear that the sex of Caliope is never questioned by herself or any of the adult characters, with the exception of her grandmother's mystical spoon test which predicted Cal would be born a boy. As the novel progresses, Cal's sex is questioned only after a trip to the emergency room for an unrelated accident.

The central question in *Middlesex* is: What role does the body play in shaping reality for the individual? Cal is raised as a girl, and the failure of his chromosomal sex to hinder his acquisition of a feminine gender schema supports the constructionist view that gender identity is completely reliant on social externalities in its development. But after Caliope finds out she could have been a boy, her personal identity becomes clear to her and she rejects the feminine role in which she was reared. Cal's ultimate rejection of feminization prior to learning any masculine traits, however, supports the essentialist ideology of identity. This rejection of an identity constructed around the corporeal body by culture for an unrealized internalized identity is voiced by Cal when he says, ". . . I never felt out of place being a girl, I still don't feel entirely at home among men" (Eugenides 2002: 479).

Identification of sex is a paradox within the novel because the characters view sex as unquestioningly determined by the external appearance of genitalia until Cal's situation is revealed and the values used in classification lose their solidity. Towards the end of the novel, the author discusses the difficulty humanity has faced in the past in constructing a universal classification system for sex. The most definitive answer reached, according to the Eugenides, was that of Klebs system created in 1876 that identified sex according to gonad tissue (2002: 410). However, even with Klebs and others using a variety of methods to identify sex, the predominant method has remained the external appearance of genitalia. Even vaguer is that sex identification within social interaction is based primarily on cultural artifacts such as clothing and hair length. The fact that Cal undergoes such a dramatic struggle when he learns he is not a *real* girl indicates the importance of sex within a social structure. Furthermore, the fact that Cal has to learn a completely different structure of daily life after he begins to live as a man signifies the social importance that the body be identified as male or female at birth because sex determines the social reality the child is raised in.

Near the end of the novel, Cal states the change in his body does not alter who he truly is. "My change from girl to boy was far less dramatic than the distance anybody travels from infancy to adulthood. In most ways I remained the person I'd always been" (Eugenides 2002: 520). This statement corresponds with the needs of transgender and intersex individuals to identify themselves within lesbian communities. The label of lesbian performs the same social function for transgender people in lesbian communities as does the label of sex for Cal in general society. The labels validate the individual's identity in terms of social interaction and perception but are not actually products of the individual's identity construction. In other words, Cal did not need to grow up being called a male in order to see himself as a male. The male label for Cal only provides the external social facet needed for him to associate with the male community.

GENDER SCHEMA ACQUISITION AS INNATE OR EXTERNAL

The concept that gender identity has an innate component is supported through some of Cal's statements, such as ". . . and yet, invisibly but unmistakably, I began to exude some kind of masculinity, in the way I tossed up and caught my eraser, for instance" (Eugenides 2002: 304). Furthermore, looking back on his childhood, Cal admits some of his tastes were masculine in nature, for "Maybe this was another sign of the hormones manifesting themselves silently inside me. For while my classmates found *The Iliad* too bloody for their taste. . . . I thrilled to the stabbings and beheadings, the gouging out of eyes, the juicy eviscerations" (Eugenides 2002: 322). Cal even viewed the dynamics of interaction between the sexes from a transcendental viewpoint. "Did I see through the male tricks because I was destined to scheme that way myself? Or do girls see through the tricks, too, and just pretend not to notice" (Eugenides 2002: 371).

Opposing Cal's personal deconstruction of gender identity is the concept that gender is created through socialization during infancy and childhood. In the middle of the novel, Cal describes a home movie taken at Easter 1962. In the movie Tessie gives Cal a doll that Cal then proceeds to nurse with a bottle. Dr. Luce uses the film to indicate Cal acquired a feminine gender identity because Cal's actions are stereotypically feminine. In the interviews between Dr. Luce and Cal, Dr. Luce itemizes Cal's external behaviors and classifies them as feminine, allowing him to conclude other than the nonconformity of his body, Cal is a girl. "He paid attention to the way I coughed, laughed, scratched my head, spoke; in sum, all the external manifestations of what he called my gender identity" (Eugenides 2002: 408). Combining his study of Cal with his other patients, Dr. Luce presents the idea that gender identity is learned and imprinted on the individual in childhood (Eugenides 2002: 411).

Cal's struggle at achieving a culturally acceptable gender identity highlights the strict demands of sex-linked gender ideology. The failure of his body to conform to society's expectations of a male body leads Cal to place a greater emphasis on masculine behaviors as a form of compensation. Cal's actions then cease to be truly representative of his self, in much the same way some transsexuals assume certain behaviors in order to pass within a transsexual community as discussed by Mason-Schrock. It is the internalized failure of knowing he can never be a *real man* that drives Cal to hide within stereotypical masculine behaviors. "I leave and never call them again. Just like a guy" (Eugenides 2002: 107). Cal's *failure* poses the risk exclusionary lesbian communities face by creating a lesbian body where even *womyn-born womyn* lesbians might feel as if they do not have lesbian bodies.

When Cal is placed within the spherical characterization model the entire concept of a gender identity becomes moot and the behaviors Cal acquires simply become part of his unique identity. The corporeality of Cal's body only affects his identity in terms of the social reactions to his appearance, but under the spherical characterization model, he does not have to give up any of the behaviors he learned in his childhood to claim a male identity. Within the spherical characterization model, Cal is able to retain his male identity without feeling the need to compensate with stereotypical behaviors that are outside of his character.

For intersex individuals, this signifies their inclusion within lesbian communities because their acquired identity is based on the Body and not the social meanings attached to the corporeal body. Cal's acquisition of behaviors around the external label of his sex illustrates the malleability of corporeality in terms of applied social meanings on identity. Cal's experience illustrates Bornstein's statement that identity is not a static collection of behaviors applied to an individual but a fluidity of behaviors the individual uses his Body to navigate through.

CONCLUSION

Inclusion of transgender and intersex individuals within lesbian communities can occur by quantifying the lesbian body to encompass transgender and intersex bodies, but creating any definitive body for a group identity leads to the development of an excluded identity. Underneath an essentialist or constructionist ideology, the excluded identity is quantified arbitrarily as is the ideal body of the group. By framing the corporeal body and mental self within the Body, the spherical characterization model allows for a community identity to develop while mitigating the circumstances for exclusion.

Both Cal and Orlando demonstrate instances for transgender and intersex inclusion in lesbian communities. Their narratives reveal the struggles individuals endure not only when their Bodies fall outside of the cultural binary but also when they find their identities excluded from social discourse. Cal and Orlando's inclusion in lesbian communities could be argued based on sexuality: Orlando's sexuality changes with her body suggesting that while she maintains heterosexual relationships, her mixed experience means she is engaging in homosexual affairs, and Cal grows up believing he is a girl who is attracted to other girls, a representation of the essential experience lesbian feminists based their exclusionary arguments around. But inclusion based on quantifiable standards inherently engenders hierarchies both within and outside of a community; under the Spherical Characterization Model the categories of heterosexual and homosexual become moot.

The Spherical Characterization Model eliminates the need for having a group *check-list* to determine inclusion (a good example of an identity check-list would be the recitation of medical texts by early transsexuals in order to qualify for surgery). Within the model, the only point a Body needs to connect with in order to belong to a group is the one of that identity. Technically, anyone can claim a lesbian identity through the spherical characterization model. This loose criteria for inclusion seems to threaten the value of having a group identity, but it is the social value of the community interaction that will determine who affiliates themselves with the group identity. The social principles of a lesbian community can remain specific because they are encoded as an identity point on the sphere; however, group principles cannot be based on quantifiable essentialist or constructionist elements, they must be based on the socialization of the group removed from the Body. What cannot remain specific is what Bodies can be included since the spherical characterization model holds that every Body retains the ability to connect to each point on the sphere. Therefore, inclusion in lesbian communities comes through the unrestricted application of that identity by the Body, signifying that both transgender and intersex individuals are justifiably included in lesbian communities.

ISSUES AND QUESTIONS FOR FURTHER CONSIDERATION

1. Positing identity as a malleable construction centered around personal choices, how can the Spherical Characterization Model be used to view traits considered biologically determined?

2. Is the Spherical Characterization Model applicable in examining the development of identities in children?

3. Does exploration of the conscious corporeal alterations made by transsexual and intersex individuals reveal that the importance of an individuals body is largely personal?

NOTES

1. For a summary of the term abject see Butler page 243.

2. For a summary of the debate on the Michigan Womyn's Music Festival see Boyd pages 134-52.

3. The term model is used instead of ideology because this paradigm is meant to be a 3-dimensional understanding of identity instead of a method for social classification; therefore the model is a new method of discourse. A basic dichotic ideology is visually represented as two separate spheres or opposite points on a line. Visualizing the identity within a dichotic system results in an individual being associated with all the behaviors ascribed to the identity as a group. To break free of grouping mentality, the spherical characterization model is visualized as a hollow 3-dimensional sphere with one point in the center that represents the individual and an infinite amount of points on the sphere surface to represent behavior and appearance.

4. The capital Body represents the visualization of my concept that identity is a singular structure of both the corporeal body and mental self.

5. Sex-linked gender ideology is how I refer to the interconnection in contemporary American society of gender and sex being linked through the cultural expectations that the sexes act and think differently as the result of innate biological differences between maleness and femaleness.

6. This represents a common justification by the medical establishment. However, infant surgeries are performed as much to allay the anxiety of parents as they are to help the child to live a "normal" (heterosexual) life. For a more in depth analysis see Dreger AD (Ed.) (1999). *Intersex in the age of ethics*. Frederick, MD: University Press Group; and Kessler SA (1998). *Lessons from the intersexed*. New Brunswick, NJ: Rutgers University Press.

7. This holds true even under argument that all women are born into oppression and therefore share a common experience as slaves under patriarchal rule. While the oppression may stem from a common source, each woman's experience under the oppression is unique.

8. Non-op transsexuals are individuals that live in the sex opposite to that which was assigned at birth but opt not to undergo surgical transformation. (Re)constructed intersex refers to individuals assigned as female or male at birth by medical professionals and underwent surgery (without their consent) to confirm this assignment. At some point later in life, the assigned sex is rejected in lieu of adopting an opposite-sex or gender variant identity, with (or without) additional surgery or hormonal treatments–in this case based on individual choice and consent.

9. I use transcend to signify the state of cognition that people obtain when they are removed from typical social participation and forced to scrutinize social interaction from the outside. This technique in changing perception leads to a transcendental observer. Boyd's discussion of Sky and Mike, two female-to-male transgender individu-

als, illustrates how this transcendental state is obtained. Sky and Mike feel their identities are male but at the same time value their prior socialization as women (Boyd 146). Sky and Mike reveal that although the social perception of them may be male, their internal social perception is a mixture of their chosen male identities with their raised female identities. The process of transcendental observation is also employed in literature as the role assumed by the reader.

REFERENCES

Bem S (1993). *The Lenses of Gender*. New Haven, CT: Yale University Press.

Boyd NA (1997). Bodies in Motion: Lesbian and Transsexual Histories. In: Duberman M (Ed.), *A Queer World: The Center for Lesbian and Gay Studies Reader*. New York: New York University Press, pgs. 134-52.

Bornstein, K. (1995). *Gender Outlaw: On Men, Women, and the Rest of Us*. New York: Vintage Books.

Butler J (1993). *Bodies That Matter*. New York: Routledge.

Eugenides J (2002). *Middlesex*. New York: Farrar, Straus and Giroux.

Gatens M (1996). *Imaginary Bodies: Ethics, Power and Corporeality*. New York: Routledge.

Lacquer T (1990). *Making Sex: Body and Gender from Greeks to Freud*. Cambridge, MA: Harvard University Press.

Mason-Schrock D (1996). Transsexuals' Narrative Construction of the "True Self". *Social Psychology Quarterly, 59* (3), 176-92.

Turner SS (1999). Intersex Identities: Locating New Intersections of Sex and Gender. *Gender and Society, 13*(4), 457-79.

Woolf V (1956). *Orlando: A Biography*. New York: Harcourt.

An *Other* Space:
Between and Beyond Lesbian-Normativity and Trans-Normativity

Myfanwy McDonald

SUMMARY. What happens when sectors of a lesbian community advocate for a space free of trans women? How do the responses to those types of policies, from within MTF trans communities, impact upon MTF trans women as a whole? How do these types of conflicts impact upon lesbians whose relationship to the term 'born female' is not straightforward? What does the focus upon these instances of conflict in the mainstream media mean and what effect does this focus have upon broader understandings about lesbians, lesbian communities and the complex intersections between lesbian and trans? *[Article copies available for a fee from The Haworth Document Delivery Service: 1-800-HAWORTH. E-mail address: <docdelivery@haworthpress.com> Website: <http://www.HaworthPress.com> © 2006 by The Haworth Press, Inc. All rights reserved.]*

Myfanwy McDonald has a BA (Hons)/BCA from the University of Wollongong in Australia and is currently undertaking a PhD at the Centre for Women's Studies and Gender Research at Monash University in Melbourne.

Address correspondence to: Myfanwy McDonald, Centre for Women's Studies and Gender Research, School of Political and Social Inquiry, Monash University, Victoria 3800, Australia (E-mail: msmcd1@student.monash.edu.au).

[Haworth co-indexing entry note]: "An *Other* Space: Between and Beyond Lesbian-Normativity and Trans-Normativity." McDonald, Myfanwy. Co-published simultaneously in *Journal of Lesbian Studies* (Harrington Park Press, an imprint of The Haworth Press, Inc.) Vol. 10, No. 1/2, 2006, pp. 201-214; and: *Challenging Lesbian Norms: Intersex, Transgender, Intersectional, and Queer Perspectives* (ed: Angela Pattatucci Aragón) Harrington Park Press, an imprint of The Haworth Press, Inc., 2006, pp. 201-214. Single or multiple copies of this article are available for a fee from The Haworth Document Delivery Service [1-800-HAWORTH, 9:00 a.m. - 5:00 p.m. (EST). E-mail address: docdelivery@haworthpress.com].

doi:10.1300/J155v10n01_10

KEYWORDS. Lesbian normativity, trans nomativity, Australian lesbian communities, Australian trans communities, lesbian social research

Lesbians born female have been denied the right to gather. . . . We intend to work for legal and community recognition, however at the moment, due to the continuing threat of legal action against us by some transsexuals and transgenders, the festival has been cancelled. (Holland-Moore, 2003)

In January 2004 Lesfest, a lesbian festival, was due to take place in the southeastern Australian state of Victoria. Prior to the event Lesfest was attracting attention within both Australian GLBTI communities and the broader Australian public for its controversial policy of excluding anyone from participation in the festival who was not a "lesbian born female" (Baker, 2003). Policies of trans exclusion in lesbian spaces in Australia have a troubled history. In the early 1990s the Lesbian Space Project, a project aimed at setting up a women-only community-based building in Sydney, faced similar problems because of conflicts over who should be allowed access to "lesbian space." Trans women were a central focus in those debates (Fox, 2002).[1]

General responses to the Lesfest policy of trans-exclusion were negative. For example, a satirical cartoon in a popular Victorian gay and lesbian newspaper showed a woman in a white laboratory coat outside the Lesfest gates looking between the legs of a woman with a magnifying glass. The caption read: "Lorraine Presents Her Credentials."[2] The Australian WOMAN Network, a group representing a certain sector within the MTF transsexual community, responded to the "lesbian born female" policy by arguing that transsexual women were legally women and that the policy was discriminatory (Carr, 2003).

Originally Lesfest was granted an exemption from Victorian equal opportunity laws that restrict these types of exclusive practices (Bedford, 2003; Baker, 2003). However, the initial decision was revoked after the Victorian Civil and Administrative Appeals Tribunal ruled that the policy *was* in fact discriminatory (Carr, 2003). Subsequently, the organisers of the festival, in an extraordinary move, cancelled the event altogether. The reason offered for the cancellation was that they "could not accept a decision that negated their right to hold an event restricted to lesbians born female" (Carr, 2003).

The Lesfest organisers' attempts to restrict access to the festival based upon whether one is a "lesbian born female" clearly ostracises a group of people, a group including not only transsexual and transgender women but also certain women with intersex conditions.[3] These restrictions constitute an example of "othering," a process that Schor describes as "attributing to the objectified Other a difference that serves to legitimate her oppression."[4] In this case the Lesfest organisers are attributing a group of women the difference of not being born female which thereby legitimates their exclusion from the Lesfest event.

Another perhaps less obvious form of othering emerges from the *opposition* to the Lesfest policy by organisations like the Australian WOMAN Network. The Australian WOMAN Network, in arguing that transsexual women should be allowed access to the Lesfest event because they are "legally women," relegates those trans women whose gender identity has not received a legal stamp of approval to an "other" status. At the present time different states in Australia have different laws regarding who should have the right to legally change their birth certificate and thereby be viewed as "legally women" (MTF) or "legally men" (FTM) (Sharpe, 2002: 160-163). However, at the time the Lesfest debates were occurring, the Victorian government had not reached a decision regarding this legislation (Baker, 2004).

Interestingly, as well as opposing the exclusive policy of Lesfest, the Australian WOMAN Network has also opposed the right for people who have not undergone MTF surgical sex reassignment to change their birth certificates (Australian W-O-M-A-N Network, 2004).[5] Thus, the Australian WOMAN Network is instituting another form of othering. That is, whilst Lesfest organisers are othering women who are not born female (thus producing a kind of lesbian-normativity[6]), the Australian WOMAN Network is othering women who have not had surgical sex reassignment (thus producing a kind of "trans-normativity").

Concurrent with the media attention surrounding Lesfest I was conducting interviews with 18 people (13 MTF and 5 FTM) who had either undergone or were thinking about undergoing some type of sex reassignment procedure (surgical and/or hormonal). These interviews were part of a three-year independent research project investigating power dynamics between physicians and their patients.[7]

In this article I use data obtained from the interviews and draw upon my own experiences as a lesbian researcher to investigate the *other* spaces produced by the debates surrounding the Lesfest event. As discussed, the *Lesfest* stance creates a kind of lesbian-normativity. Lesbian-normativity in this case is determined by an arbitrary born female vs. not-born female binary. The Australian WOMAN Network stance creates a kind of trans-normativity determined by the law. Trans-normativity relies upon a subjective legal woman vs. "illegal woman" binary. The "other" spaces I investigate in this article lie somewhere in-between or beyond lesbian-normativity and trans-normativity.

The "other" spaces are not only spaces inhabited by bodies; they also are spaces where certain stories about being a lesbian and/or being trans are told. These spaces complicate the seemingly simplistic oppositional model of trans/lesbian interaction in publicised debates about events like Lesfest. In the first section of the article I investigate the "other" space of MTF trans women who don't have MTF bottom sex reassignment surgery (i.e., surgical sex reassignment on the genitals). In the second section I investigate the "other" story of a non-trans lesbian who has considered FTM sex reassignment. In the conclusion I discuss some of the implications of the mainstream media's focus on events like "Lesfest" where trans-normativity and lesbian-normativity are potentially viewed as representations of trans/lesbian interactions as a whole.

"ILLEGAL" MTF LESBIAN TRANS WOMEN

In the publicised debates about Lesfest, and in debates about similar exclusive lesbian events in other Western countries such as the U.S.A., it often seems that the categories transsexual/transgender and lesbian are oppositional, antithetical, like magnets that repel–unable to coexist. This is evidenced in the Lesfest debates. The organisers concluded that Lesfest literally could not exist as an event that included transsexual/ transgender patrons. In these publicised controversies the Lesfest organisers and organisations like the Australian WOMAN Network become public representations of two supposedly distinct communities with two assumed oppositional points of view. However, in the process of conducting my research project, I discovered a more complex, multi-layered relationship between transsexuals/transgenders and lesbians.

The MTF trans lesbians I interviewed (nine out of the thirteen self-identified as lesbians) did not express a sense of not belonging with local lesbian communities. Neither did they express a sense of exclusion nor, indeed, a fear of exclusion from those communities. On the contrary, not belonging to, or having a feeling of or alienation from, local trans communities, for almost all the MTF lesbians interviewed, seemed to be a much more pressing issue. This was an interesting finding as it highlighted some of the complex conflicts within MTF trans communities and challenged the idea that the Australian WOMAN Network is representative of the Australian MTF trans community as a whole.

To illustrate, Sylvia, an MTF woman in her late 30s identifying as transgendered, stated that she "feels comfortable" as part of the local GLBTI community; however, she also described how she faced criticism within a certain sector of the local MTF trans community because she had decided not to have MTF bottom surgery. Sylvia describes in this extract those responses to her decision and how some trans women claimed that she wasn't transsexual:

> [The responses were] a bit judgemental. A couple of people said you are *not* a transsexual but [I said] look I was going nuts trying to live as a male [and they said] no you are not a transsexual. They were quite blunt about it one person said you're not a transsexual [because] you haven't transitioned.

In this instance Sylvia becomes the "objectified other." She is literally excluded from the category transsexual because she doesn't conform to a normative model of what the MTF transsexual is within this group setting—a person who has undergone bottom surgery. Sylvia challenges this model of trans-normativity and argues that:

> Transsexual covers anyone who lives full time in the physical gender opposite to what they were assigned at birth, whether they're taking hormones [or] considering surgery.

Time and again this issue of not belonging to, or having a feeling of alienation from, local trans communities, compared to a general feeling of acceptance within lesbian communities, arose in my interviews with MTF trans lesbians. Clearly Sylvia exists outside the boundaries of trans-normativity because she doesn't believe that the only *true* transsexual is the transsexual who has had sex reassignment surgery. However some of the trans women I interviewed existed outside this boundary in more complex ways.

Danielle is a 50-year-old lesbian who self-describes as a "man-made tranny." She states that she's "had no trouble at all" being accepted in lesbian communities. However, she asserts that she doesn't

> [H]ave a lot in common with transsexuals in the sense that they're looking to become a woman. . . . I don't know what that means. . . . I went along to a transsexual therapy group and had a lot of diffi-

culty [with the way] they were trying to achieve some sort of integration with a paradigm.

Danielle's sense of alienation within certain MTF trans groups is perhaps partly due to her status as a person who was born with an intersex condition. That is, although she identifies as a "tranny," Danielle's history differs quite significantly from many other MTF transsexual women in that she underwent an operation as an infant to "correct" her so-called "ambiguous genitalia." This surgical intervention during infancy provides the basis for Danielle's self-described status as a "man-made tranny." Danielle states that she has never felt "particularly XY or particularly anything."

Danielle exists outside the boundaries of trans-normativity in several ways. She *has* had bottom surgery, but in her case it occurred during infancy in order to achieve alignment between anatomy and sex assignment. At infancy she went from being "ambiguous" to being assigned a boy. At the time I interviewed Danielle, she was waiting to undergo an operation that would, from her perspective, "undo what they did." Although the medical establishment would characterize Danielle as a person awaiting MTF surgery, this represents neither her material reality nor how she perceives the process.

Clearly Danielle's transition is not as straightforward as many other MTF women. She was not "born female" or at least did not receive this assignment at birth. Thus, she does not fit the ideals of lesbian-normativity. But she also does not fit the ideals of trans-normativity because she has not yet undergone surgical MTF sex reassignment and she does not feel as if she is "particularly anything." Like Sylvia, she exists in an "other" space.

One of the problematic aspects of othering is that it denies the complex differences that exist amongst the group that is othered. The Lesfest organisers, by excluding all women who weren't born female, undermine the complex differences within that group of women not born female.[8] So, for example, trans women like Sylvia, whose perspective is strongly influenced by transgender theory and politics (a body of thought that challenges trans people's reliance upon the institution of medicine) and Danielle, whose experiences differ significantly from many trans women who don't have intersex histories, become part of a presumed homogenous group–those women not born female.

The Australian WOMAN Network, in arguing that the only trans women who are "legally" women are those who have undergone surgical sex reassignment on their genitalia, also undermines the differences amongst those trans women who haven't had surgical sex reassignment. That is, the reasons *why* trans women occupy an Other space differ significantly. For example, Sylvia has *chosen* to occupy that space, as a trans woman who chooses not to have surgery. However some of the other trans women I interviewed were unable to afford the type of surgery that would have enabled them to occupy the site of the "legal" woman because, in the majority of cases, surgical sex reassignment is not covered by public health insurance in Australia.[9] This is one of the most worrisome aspects associated with conceptualising MTF trans women that have not had surgical sex reassignment as "illegal." If the "legal" woman is the one who has had surgical sex reassignment, then many trans women on low incomes who can't afford the surgery will *unwillingly* inhabit the other space of the "illegal" woman. They will be doubly excluded–as low-income earners in the general community and as "illegal" women under the law and within certain sectors of trans communities. Hence, identity becomes a function of capitalistic privilege.

Sylvia's and Danielle's stories exemplify some of the diverse range of experiences in MTF trans communities that are either overlooked or marginalized via the production of lesbian-normativity and trans-normativity. As stated previously, through my research I discovered a complex, multi-layered relationship between trans and lesbian–a relationship that was being ignored and/or denied in the publicised debates surrounding Lesfest. This discovery was intimately tied to an exploration of my own location as a lesbian researcher in this project.

THE OTHER STORIES OF LESBIANS BORN FEMALE

Nathan: I can do an interview, although I'd like to ask beforehand, are you trans yourself?

Myfanwy: No, I have not had any sex reassignment procedures, hormonal or surgical. Although I have thought about having SRS [and that] is one of the motivations for this research.

The above quote comes from an e-mail conversation I had with Nathan, an FTM trans man in his mid-twenties self-describing as a "tranny-boy." Throughout the process of conducting my research I was open about the fact that I was a lesbian who was not trans but had considered FTM sex reassignment. If the research participants asked, as

Nathan did, I would share my experience. The process whereby I came to consider and narrate my own story provides an insight into the "other" space that I have described. This space complicates the seemingly simplistic oppositional model of trans/lesbian interaction in publicised debates about events like Lesfest.

My decision to be open about my location, as a non-trans lesbian who had considered FTM sex reassignment, was initially related to my desire, as a social researcher, not to position myself as an objective, neutral observer. Many feminist, gay, lesbian, queer and Black theorists have noted the problems with adopting such a stance (see, for example, Game, 1991; Collins, 1991; Honeychurch, 1996). In short, I wanted the research participants to know that I had a stake in this research, as someone who had considered FTM sex reassignment. As the Lesfest debates emerged, however, this openness seemed even more important, as I needed to make clear the complex nature of my status as "lesbian born female."

The public statements from the Lesfest organisers tended to suggest that femaleness was something that needed to be recognised, protected, and fostered. However, as I state to Nathan in the above quote, I have often wondered if I should have FTM sex reassignment. I am, intermittently, extremely uncomfortable with female embodiment. In that sense femaleness is not something I have experienced as an essence that needs to be recognised, protected, and fostered. On the contrary, it is something I intermittently want to ignore and perhaps even shed.

Clearly I exist well outside the boundaries of trans-normativity in that I haven't had any type of sex reassignment procedure. However, I am *not* an other to lesbian-normativity in that I *was* born female. In this sense I am *not* an other as Sylvia and Danielle are. I am not physically embodying an other space as my body is firmly placed within the lesbian-normative location of female. My story however exists in an other space, even if physically I do not. This is a story that overlaps between trans and lesbian or exists somewhere in between. It is the story of a woman who is attracted to women but a woman who, intermittently, longs for FTM sex reassignment. The oppositional trans/lesbian model, however, makes this story difficult to tell.

As an illustration, the publicity surrounding events occurring in Australia like the Lesfest debates creates, in my mind, a sense of urgency regarding the decision to have or not to have sex reassignment. The oppositional trans/lesbian model surrounding the Lesfest event tends to imply that one's affiliation with the trans identity, the trans community and/or trans politics determines one's allegiance or commitment to certain lesbian communities. This oppositional model, in other words, implies that one has to make a decision *between* lesbian and trans. A space in between tends to bring one's commitment to both communities into question. That is, if you're thinking about having FTM sex reassignment you're not recognising, fostering, and protecting femaleness and if you haven't had sex reassignment you're not *truly* trans.

Obviously, this sense of urgency is no doubt linked to the embodied discomfort that feelings of gender dysphoria can have. I don't want to deny that sense of discomfort. However, I would like to explore here some of the broader historical and cultural meanings of these feelings of urgency. I would argue that for the lesbian considering FTM sex reassignment this sense of urgency is linked to the history of the development of gay and lesbian communities and politics. Before the birth of the contemporary gay and lesbian movement (occurring in the West in the late 1960s to early 1970s) lesbians were often seen as women who had the internalised sexuality of heterosexual men.[10] Altman (2001) states that at this time: "the dominant understanding of homosexuality was predicated on a confusion between sexuality and gender" (25).

The argument that being a lesbian is different from wanting to be a man when one inhabits a female body (which is one way of describing the FTM experience) has been an important part of gay and lesbian politics. It has given lesbians and gay men a standpoint from which to challenge the heteronormative expectation that the feminine/woman/female will desire and be attractive only to the masculine/man/male and vice versa (Pringle, 1992: 81-83).[11] It could also be said that this distinction made it possible for an FTM trans identity to be acknowledged.[12] In other words, this meant also that a female-bodied person who wants to be a man was not "confused" with a lesbian.[13]

Yet the lesbian who is considering FTM sex reassignment exists in a tenuous location because she carries the weight of the historical "confusion" to which Altman refers. In other words, the lesbian who is considering FTM sex reassignment is open to the charge that they are internalising the heteronormative ideal that only the male/man/masculine can desire and be attractive to the female/woman/feminine. The medical term for this (which is somewhat outdated now) is "egodystonic homosexuality." The layman's term is "internalised homophobia."

Having almost completed my research project I still have not undergone any hormonal or surgical sex reassignment procedures and I still remain undecided. In fact, I have no doubt that in the future I will continue to think about having FTM sex reassignment. I recognise that one of the dangers of staking a claim on an in between space is that the unique experiences of people who have undergone sex reassignment and do identify as trans could be diluted. Obviously thinking about undergoing sex reassignment and actually undergoing sex reassignment, with all the accompanying legal and social difficulties associated with that transition, are two very different things. However, what I am interested in here is the possibility of telling the story of an "other" space.

As stated previously, in the publicised debates about Lesfest, and in debates about similar exclusive lesbian events, it often seems that the categories transsexual/transgender and lesbian are oppositional, antithetical, like magnets that repel–unable to coexist. This oppositional discourse leaves little space for a story that is neither exclusively the story of the woman who is attracted to women nor only the story of the woman who thinks they might want to be/live/identify as a man. Indeed, this discourse leaves the undecided lesbian in an especially tenuous space because of the potential charge of internalised homophobia or egodystonic homosexuality.

In saying this I am not alleging that lesbians who consider FTM sex reassignment are somehow routinely excluded from lesbian spaces in Australia. On the contrary, the diversity of lesbian communities in Australia allows for a range of different discussions, possibilities, and identities to flourish. What I am saying, however, is that in the *publicised* debates surrounding events like Lesfest this diversity and complexity is overlooked and undermined. The complex, multi-layered relationship between transsexual/transgender and lesbian is condensed and packaged as a catfight between two seemingly uncompromising, exclusive, and distinct communities. This representation has a problematic impact upon those individuals who experience, embody, or tell a story that does not conform to this oppositional model.

CONCLUSIONS

What I set out to do in this article was to investigate the "other" spaces that complicate the seemingly simplistic oppositional model of trans/lesbian interaction in publicised debates about events like Lesfest It is clear now how the production of an "other" via lesbian-normativity and trans-normativity can have problematic impacts upon understandings of lesbian and trans communities and the lives of the individuals who inhabit "other" spaces or tell "other" stories.

In Australia the issue at this stage is not how to make lesbian communities, lesbian spaces or lesbian events *as a whole* more inclusive for trans women. After all, according to the MTF trans women who I interviewed, lesbian communities and spaces in Australia generally *do* seem to be inclusive of trans women. Clearly the problematic policies of events like Lesfest cannot be ignored. However, it is important to remember that these policies are not representative of Australian lesbian communities in general. Indeed, one issue that remains inadequately explored is why the mainstream gay and lesbian media and the mainstream media in general focus so intensely upon those moments when certain factions of the lesbian community and certain factions of the trans community are in conflict.

One of the perplexing issues I faced whilst conducting interviews for my research was why, if MTF trans lesbians don't feel excluded from lesbian communities but do feel alienated within certain sectors of MTF trans communities, is so much focus placed upon the trans-exclusive policies of Lesfest organisers? Why aren't the feelings of alienation that some MTF trans women feel in some MTF trans communities also a public issue? Clearly the Lesfest debate attracted attention because it was an overt conflict occurring in a public arena. It was, in a colloquial sense, "hot gossip." In that sense, it could be argued, the focus on Lesfest is predictable and not especially worrying.

I would argue, however, that there is a problematic side to this focus on Lesfest. The controversy surrounding the event 'makes sense' in the Australian mainstream media because it reinforces the stereotype of the hard-line, uncompromising lesbian. On the other hand, a diverse lesbian community that allows for a range of different discussions, possibilities, and identities to flourish, challenges that stereotype. I am not alleging that the mainstream media is consciously not choosing to cover stories that depict inclusive lesbian communities. Rather, I mean to suggest that certain stories about lesbian communities confirm stereotypes held by the general reading audience. I wonder if, instead of criticising certain sectors of lesbian communities for being trans-exclusive, we should question why the most publicised stories that could be told about the relationship between trans and lesbian are (a) the stories of the catfight and (b) the stories where trans and lesbian exist as polar opposites.

This brings me to another problematic issue that emerges from the focus on the Lesfest debate within the mainstream gay and lesbian media, considering the range of other stories that could be told. For those non-trans lesbians who read mainstream gay and lesbian publications and don't know about the diversity of Australian trans communities, the Australian WOMAN Network could easily be viewed as representative of Australian trans communities as a whole. This potentially could increase antagonism towards trans women within lesbian communities, an antagonism that, at least according to the MTF trans lesbians I interviewed, is *not* currently prevalent. For example, in 2003 one contributor to an online discussion on a lesbian Website about the Lesfest debates stated:

> I have always supported diversity within lesbianism, but with the actions of the Australian WOMAN Network I withdraw any support for transsexuals in our community. (Pink Sofa, 2003)

If the stories of the Other (the stories existing outside the oppositional trans/lesbian model) were more widely acknowledged and publicised, there is a greater likelihood that non-trans lesbians could see the perspective of the Australian WOMAN Network for what it is–*one* perspective in a diverse range of trans communities that contain within them many different, in some cases opposing, perspectives. These diverse trans communities, and the individuals who are linked to them, intersect in complex ways with diverse lesbian communities. There is no single way of representing these multitudinous intersections. The debates surrounding Lesfest tell one story. But there are *other* stories in Australia that can be told in order capture the complex and evolving nature of the relationship between trans and lesbian.

AUTHOR NOTE

Thanks to Dr. Maryanne Dever and Dr. Rachel Fensham for their continued support; thanks to Angela for her invaluable help with all the drafts; thanks to Sylvia, Danielle, Nathan and all other the participants who agreed to take part in this project; special thanks to everyone involved in the 2004 NOV International Postgraduate Program at Utrecht University: "I wanted to say with regard to ideas there is no revolutionary and individual, only the evolutionary and collective" (Stanley, 1993, p. 200).

NOTES

1. The debates over trans exclusion in public lesbian spaces in Australia are especially interesting given the central role male-only bars and clubs played in the development of the Australian feminist movement. The right for women to enter those male-only venues was a pivotal turning point in the early years of the second wave of the feminist movement in Australia.

2. The cartoon appeared in the *Melbourne Community Voice* Newspaper on September 19, 2003.

3. It is difficult to concisely describe the difference between transsexual and transgender. However, transsexual is usually viewed as a term that emerges from medicalised discourse, whilst transgender is viewed as a term that emerged from a grassroots social movement that is based upon a radical approach to gender transgression (Roen, 2002). Intersex, on the other hand, is a physiological condition that is often, although not always, visible at birth where the infant is born with what is termed by the medical establishment as "ambiguous genitalia." At present in Australia the approach to the treatment of infant intersex conditions varies according to the hospital where the surgery is being done and the specific physicians involved. Prior to the mid-1970s Australian physicians routinely operated on intersex infants based purely upon the appearance of the infant's genitalia. Since the mid-1970s many Australian physicians involved in the treatment of these conditions have advocated for a more collaborative approach, taking into account the potential physical and psychological impacts of surgery, with the involvement of the parents and a range of other health professionals (Warne, 2003). However the ethical issues relating to surgery upon infants have still not been settled with intersex advocacy groups challenging some aspects of intersex surgery (see, for example, the AIS Support Group Australia, 2004).

4. See Andermahr S, Lovell T, and Wolkowitz C (2000), *A Glossary of Feminist Theory*. London: Arnold Publishing, p 157.

5. An exception is made here for "those who are medically deemed unable to undergo" sex reassignment procedures (Australian W-O-M-A-N Network, 2004).

6. See Angela Pattatucci Aragón's definition of lesbian *homonormativity* in the Introduction to this volume.

7. The interviews were primarily conducted face-to-face with people living the south-eastern Australian states of NSW and Victoria–where the Lesfest debates were receiving the most attention.

8. This practice also denies the complex differences and diversity among those women qualifying as "lesbians born female."

9. Australia's public health insurance scheme, called Medicare, covers most of the costs of surgical procedures considered to be "essential." Procedures deemed "elective" are not covered by Medicare. Surgical sex reassignment is, in most cases, deemed "elective" in Australia. There seems to be no set cost for MTF genital surgical sex reassignment; therefore, it is difficult to estimate the exact cost. However, a person seeking that treatment can expect to pay up to $30,000 (Australian).

10. See Faderman, 1991: 41-45 for further discussion of these theories.

11. My definition of heteronormativity here is based upon Butler's (1990) discussion on the concept of "intelligible" gender, that is where gender "follows" from sex and desire "follows" from either sex or gender (17). So, under this cultural matrix, only a woman can desire a man, only the female desires the male and so on.

12. It is important to acknowledge here that the experiences of FTM trans men cannot all be described as the experience of wanting to be a man when one inhabits a female body.

13. This particular distinction is still a topic of debate in terms of claiming historical figures as either FTM trans men or lesbians. See Boyd, 1991 and Hale, 1998 for examples.

REFERENCES

AIS Support Group (2003). *Aims of The AIS Support Group Australia.* Retrieved 18 May 2005, from *http://home.vicnet.net.au/~aissg/AimsandPolicies.htm*

Altman D (2001). Rupture or Continuity: The Internationalisation of Gay Identities. In: Hawley JC (Ed.), *Postcolonial, Queer: Theoretical Intersections.* New York: State University of New York Press.

Australian W-O-M-A-N Network. (2004). *Birth Certificate Correction: Human Rights For People of Transsexual Background.* Retrieved 7 May 2005, from *www.w-o-m-a-n.net/state%20pages/victoria.htm*

Baker R (2003). *Lesbians Only Need Apply.* Retrieved 15 September, 2003, from www.theage.com.au/articles/2003/09/11/1063268514810.html

Baker R (2004). *State to Move On Transsexual Status.* Retrieved 8 March, 2003, from *www.theage.com.au/articles/2004/03/03/1078295446973*

Bedford K (2003). *Lesbians Only at Daylesford Festival.* Retrieved 15 September 2003, from *www.abc.net.au/victoria/stories/s946290.html.*

Boyd NA (1999). The Materiality of Gender: Looking for Lesbian Bodies in Transgender History. *Journal of Lesbian Studies, 3*(3), 73-81.

Butler J (1990). *Gender Trouble: Feminism and the Subversion of Identity* New York: Routledge.

Carr A (2003, October 23). Lesfest Cancelled. *BNews, 76,* 7.

Collins PH (1991). Learning From the Outsider Within. In: Funow MM and Cook JA (Eds.), *Beyond Methodology.* Bloomington: Indiana University Press.

Faderman L (1991). *Odd Girls and Twilight Lovers: A History of Lesbian Life within Twentieth-Century America.* New York: Columbia University Press.

Fox K (2003). *Genderphobia . . . Do You Have It?* Retrieved January 1 2003 from *http://www.katrinafox.com/genderlotl.htm.*

Hale CJ (1998). Consuming the Living, Dis(Re)Membering the Dead in the Butch/FTM Borderlands. *GLQ: A Journal of Gay and Lesbian Studies 4*(2), 311-348.

Holland-Moore CA and Holland-Moore A (2003). *Lesfest 2004 Cancelled.* Retrieved 17 October, 2003, from *www.nrg.com.au/~wow/*

Honeychurch KG (1996). Researching Dissident Subjectivities. *Harvard Educational Review, 66*(2), 339-355.

Ingraham C (1996). The Heterosexual Imaginary: Feminist Sociology and Theories of Gender. In: Seidman S (Ed.), *Queer Theory/Sociology.* Oxford: Blackwell.

Pink Sofa. (2003). *Grrltalk Forum.* Retrieved 12 December 2003, from http://www.thepinksofa.com.au.

Pringle R (1992). Absolute Sex? Unpacking the Sexuality/Gender Relationship. In: Connell RW and Dowsett GW (Eds.), *Rethinking Sex: Social Theory and Sexuality Research,* Melbourne, Australia: Melbourne University Press.

Roen K (2002). Either/or and Both/Neither: Discursive Tensions in Transgender Politics. *Signs 27*(2), 501-23.

Sharpe A (2002). *Transgender Jurisprudence: Dysphoric Bodies of Law.* London: Cavendish Publishing.

Stanley L (1993). Methodology Matters! In: Robinson V and Richardson D (Eds.), *Introducing Women's Studies.* London: Macmillan.

Warne G (2003). Ethical Issues in Gender Assignment. *The Endocrinologist 13*(3), 182-186.

In Another Bracket:
Trans Acceptance in Lesbian Utopia

Jamie Stuart

SUMMARY. *Better Than Chocolate* investigates the inclusion of trans-gendered women in lesbian space. The film makes a strong statement that it is logical and natural to welcome trans women to the lesbian community, and that those who reject this inclusion are old-fashioned, violent, or both. It is my assertion that the film makes this statement at the expense of both those who have nuanced reservations toward trans inclusion and those transwomen who desire inclusion into lesbian communities but aren't hyperfeminine. *[Article copies available for a fee from The Haworth Document Delivery Service: 1-800-HAWORTH. E-mail address: <docdelivery@haworthpress.com> Website: <http://www.HaworthPress. com> © 2006 by The Haworth Press, Inc. All rights reserved.]*

KEYWORDS. *Better Than Chocolate*, lesbian film, transgender characters, LGBT cinema, queer cinema, gender performance, transgender acceptance, Canadian film

Jamie Stuart is a doctoral student in American Culture Studies at Bowling Green State University. Her primary research interest is the expression of gender and sexuality in film and other media. Her dissertation focuses on stage performance in lesbian-themed films.

Address correspondence to: Jamie Stuart, Department of American Culture Studies, 101 East Hall, Bowling Green State University, Bowling Green, OH 43403 (E-mail: stuartj@bgnet.bgsu.edu).

[Haworth co-indexing entry note]: "In Another Bracket: Trans Acceptance in Lesbian Utopia." Stuart, Jamie. Co-published simultaneously in *Journal of Lesbian Studies* (Harrington Park Press, an imprint of The Haworth Press, Inc.) Vol. 10, No. 1/2, 2006, pp. 215-229; and: *Challenging Lesbian Norms: Intersex, Transgender, Intersectional, and Queer Perspectives* (ed: Angela Pattatucci Aragón) Harrington Park Press, an imprint of The Haworth Press, Inc., 2006, pp. 215-229. Single or multiple copies of this article are available for a fee from The Haworth Document Delivery Service [1-800-HAWORTH, 9:00 a.m. - 5:00 p.m. (EST). E-mail address: docdelivery@haworthpress.com].

Better Than Chocolate, the popular, successful lesbian and crossover movie of 1999, reads both as a blueprint for an idealized lesbian community and as reflective of real-life lesbian worlds. The characters are so varied along an LGBT continuum that it almost looks like the film is fearful of leaving anyone out; yet the situations they find themselves dealing with are strikingly similar to our own in the real world, even six years later. Insofar as the film reads as a kind of lesbian utopia, there are strong messages regarding the inclusion of trans women: that it is logical and natural to welcome trans women to the lesbian community, and that those who reject this inclusion are old-fashioned, violent, or both. The only outcomes for women who are less than enthusiastically welcoming to trans women in lesbian communities are a voluntary change of heart or a forcible reprimand. However, the kind of trans woman that the film suggests should be accepted unconditionally into lesbian communities is very feminine, beautiful, and charming. While the film is the first to investigate the inclusion of trans women into lesbian communities as a large part of its narrative, it forwards a particular kind of woman to the exclusion of trans women who, for whatever reason, cannot or don't care to adhere to conventional models of femininity.

Among the issues the film tackles, trans[1] inclusion is given substantial attention in the narrative, with almost every major character stating their opinion. With the large lesbian audience this film enjoyed, the film can be read almost as a proposal for how things should be. This is especially the case since the inclusion of trans women in lesbian communities is an issue for many real-life communities, and the film may have influenced the way lesbians thought about the issue in their own lives.

HOW I'D LOVE TO STEP INSIDE THAT SCREEN[2]

I do not mean to imply that the filmmakers were deliberately setting out to champion the cause of trans inclusion in lesbian communities. However, films don't exist in vacuums, and when they portray issues or events that are occurring offscreen as well, taking a stand of some sort is often inevitable. "While we might not want to argue that cinema is integrally or essentially utopian," suggests Gaines (2000), "it could be said to have a *utopianizing* effect; that is, whatever subject receiving cinematic treatment can be produced as a 'wishful landscape'" (110). The fictional Vancouver inhabited by Kim, Maggie, Lila, and Judy is just such a landscape–not a perfect world, but a better one than we live in.

"Entertainment offers the image of 'something better' to escape into," says Dyer (2002) of this utopianizing effect of film. He asserts that "the utopianism [in entertainment] is contained in the feelings it embodies. It presents . . . what utopia would feel like rather than how it would be organized" (20). This is what I argue *Better Than Chocolate* does. Its actual representation of trans inclusion is flawed, and obviously would not work as smoothly in real life as it does in the story world. But the way the characters *feel* about trans inclusion and the way that the film makes *us* feel is what is important. Writing about live theatre, Dolan (2001) suggests that we watch performances "to reach for something better, for new ideas about how to be and how to be with each other" (455). I believe that the inclusive atmosphere and happy ending of *Better Than Chocolate* present viewers with examples of how they might "be with each other" in a harmonious way.

EVERYBODY'S ENTITLED TO A LITTLE FANTASY

In 1995, Stacey wrote of *Desert Hearts*: "the lack of critical attention the film has received indicates a worrying discrepancy between lesbian audiences in general, for whom this was their favourite film, and lesbian academics, whose relative silence on the film's success is striking" (95). The same is true of *Better Than Chocolate*. The film has not received much theoretical examination to date, although it remains the favorite of many. The film is particularly important to study because of the care that went into mirroring the "real" world. Thomas (2001) suggests that to "read a film is to engage with it in all of its detail as a starting point for talking about things that matter and, in the process, to discover the common ground between the film and us" (1). Dyer (2002) makes a similar claim: "to be effective, the utopian sensibility has to take off from the real experiences of the audience" (27). Those of us who love the film cannot ignore the similarities between the story world and our own worlds, and the film's assertions about trans women in lesbian communities may have had a bigger impact on real life communities because of it.

There is evidence that the filmmaking team did want the film to mirror real lesbian communities, at least in appearance. Anne Wheeler, the director of *Better Than Chocolate*, talks a lot about the creative process of making the film in her commentary track on the DVD. For example, it is here that we learn that a lot of effort was expended trying to make the film look and feel authentic. The filmmakers went to a lot of clubs to get ideas for The Cat's Ass, a nightclub that makes up one of the film's major venues. Wheeler had several twentysomething women she called "consultants" to help her and the artistic designers make a world that a twentysomething lesbian audience would read as hip. The sex scenes were monitored for authenticity. Gellman (1999) says that during filming Wheeler "would call out to the largely gay crew: 'Could we have some lesbians over here to check this out. Does this look real?'" ("Venus," para. 8).

Many of the scenarios that the characters encounter come from the lives of real individuals in Vancouver. The censorship at the bookstore is "a scenario based on the real life struggles of Vancouver's Little Sisters bookstore," according to Gellman ("Venus," para. 2). Tony kicking Maggie and Kim out of his coffee shop for attempting to kiss came from a real event. The song that Judy sings at the nightclub is based on a song a real Vancouver transwoman sings. Most of the music used in the film is performed by Vancouver-area artists. Even the chocolates that Lila continually consumes are made by a local chocolatier.

The result is a world that is very similar to our own, dealing with the same kinds of issues, only trendier, prettier, and more fun. Kirkland (1999) says that "if coming out of the closet were really as much fun as it is for the sexually adventurous youths in *Better Than Chocolate*, then everybody would be doing it, even straight people," and, later, "real life is tougher, harsher and nastier" ("Coming out," para. 1). The story world of this film is like our world, only better.

NOTHING HERE IS PADDED

Better Than Chocolate was the first lesbian feature film to have trans inclusion as one of its major themes. While much of this article dwells on the limits of this particular representation, it is still a groundbreaking film in its positive representation of a trans lesbian participating in a lesbian community in a dynamic way. Peggy Thompson, the screenwriter, says that "the film really took a gigantic step forward with the birth of Judy. She really is the soul of the movie" ("Books into Movies," para. 15). While the representation of the inclusion of trans women in lesbian communities is problematic in this film, Judy as a character is remarkably nuanced and positive in her emotional stability and ability to navigate her world with confidence.

Played by Peter Outerbridge, Judy is presented as very feminine and ladylike. In fact, Wheeler suggests that Judy's wardrobe was deliberately constructed to mirror Lila's (Maggie's mother) to some degree. Except for red, which she wears in the first and next-to-last scenes in the film, Judy wears mostly neutral colors: usually black, but she wears tan, light grey, and light blue one time each. This symbolizes her desire to not stand out (or, conversely, to blend in) and that she is tasteful in selecting her wardrobe. Additionally, all of her outfits involve layers. The outermost layer is sometimes sheer and gauzy and sometimes more businesslike, but in all except one, the bottom layer is a soft, T-shirt-looking piece. This suggests both modesty and a desire for comfort.

During Judy's second appearance in the film, when the audience (and Kim) are introduced to her, she is wearing a gauzy black blouse with small white polka dots all over it. In a subsequent scene, wearing the same blouse, she introduces herself to Lila, and Lila expresses admiration for it, implying that the two have similar tastes in fashion. Lurie (2000) asserts that this particular pattern indicates "either mere good humor or (especially when black and white are used) a sophisticated wit, satire, and irony" (209). The fact that this is what she is wearing in more than one scene, across different venues, when being introduced both to the audience and to other characters, suggests something about Judy; that she has a sense of humor and wit. In fact, in the first of these two scenes, the characters are in the bookstore, and Judy is thinking about moving into a condominium complex called Heritage Peak. When Frances makes a flip remark, Judy wittily replies, "I'll be a Heritage Homo."

Much later in the film, Judy is wearing a gauzy white shirt with a flowered pattern on it while she stencils a flowered border in her new condo. Lurie (2000) asserts that "flower patterns . . . seem to stand for femininity" (210). Lila also wears gauzy shirts and floral patterns, but she and Judy are the only women in the film who do so. They also have other similarities in wardrobe (for example, they both have animal-print scarves), and Wheeler says that in one scene they are "dressed almost identically . . . they look like real girlfriends" (commentary). Lurie suggests that transsexuals "usually wear the sort of clothes that a respectable woman of their own age and station would normally wear" (260). While this statement is obviously a generalization about MTF transsexuals,[3] it describes Judy's appearance perfectly. Judy's wardrobe (and hair and makeup) is similar to Lila's. Other commonalities also exist. Lila and Judy seem to be about the same age, they are both singers, and they are both going through painful separations from their families.

There are variations in their wardrobes, however, that reflect personality differences. For example, Judy consistently looks more pulled-together and polished than Lila, and her clothes fit her better. This mirrors the narrative implication that she is more comfortable with herself and secure in her position in life than Lila.

Wardrobe is always important in a movie, but it is especially the case for transgendered characters. Clothing for transgendered people can be seen as a barrier between their bodies and the rest of the world. Warwick and Cavallaro (1998) suggest that "there is no obvious way of demarcating the body's boundaries" (xv), but they explore the extent to which clothing can function as such a boundary. They address the ways that clothing both conceals the physical body (and, therefore, the characteristics so many use to define sex) and reveals the gender of the person through societally accepted patterns. In Judy's case, her clothing both conceals her transitioning body, which some might see as male, and reveals her female gender identity through her conscious choice of feminine items. Her performance of femininity is thoughtful and self-confident; she seems to enjoy the feminine signifiers she chooses.

As a representation of a transwoman, Judy shines; however, the brand of hyperfemininity she exibits raises some troubling questions about the film and about trans inclusion in lesbian spaces. For example, it could mean that trans women are only acceptable if they are conventionally attractive and feminine, or if they can "pass." This possible meaning comes at the expense of trans women who, for any number of reasons, either cannot or don't care to perform conventional models of femininity.

Outerbridge plays Judy with generous gestures and sensitive facial expressions. In almost every scene Judy is in, she has a moment where she throws her arms wide, away from her body, almost in a welcoming gesture. The many scenes with Lila contrast their body languages: Judy's arms are open, her gestures sweeping, but Lila often crosses her arms or fiddles with the pearls around her neck. Where Lila is nervous and fidgety, Judy is calm and smooth. Since most of Judy's interactions with others involve active listening, her facial expressions are important, and Outerbridge portrays them convincingly. In addition, there is always a look of hopeful anticipation on her face as she pursues Frances romantically. In general, Judy's appearance and demeanor are more comfortable and self-assured in their expression than Lila's.

The relationship between Judy and Lila illustrates the performative nature of femininity. Judy, as a trans woman, is likely seen by some as imitating or copying *real* women. This argument, however, rejects or ignores the constructed and performative nature of all gender, and is described by Butler (1990) as "assum[ing] that there is a 'doer' behind the deed" of gender (25). Butler suggests that a "repetition of 'the origi-nal' . . . reveals the original to be nothing other than a parody of the *idea* of the natural and the original" (31, emphasis original). In this case, rather than Judy being a copy of Lila's original, her performance merely calls to attention the unoriginality of Lila's feminine performance. If anything, Lila looks up to and emulates Judy for her singing career, her choice of clothing, and her confident physical presence, which further subverts the copy/original paradigm.

While the film presents a world in which the relationship between the original and the copy is ruptured and in which a lesbian community en-thusiastically welcomes a trans woman, trans acceptance and under-standing of trans issues are still not a reality in the real world, inside and outside of lesbian communities. Dolan (2001) asserts that through per-formance "it's possible to imagine a utopia where the social scourges that currently plague us might be ameliorated, cured, redressed, solved, never to haunt us again" (456-7). *Better Than Chocolate* offers us a vi-sion of such a world. I find myself agreeing with Dolan, however, when she suggests that she has "faith in the possibility that we can imagine such a place, even though I know we can only imagine it, that we'll never achieve it in our lifetimes" (457). While most of the film's charac-ters like and accept Judy, the story world is still a utopic one that does not necessarily have offscreen authenticity.

NOT A FUCKING DRAG QUEEN

Although Judy occasionally makes reference to "the surgery" and "the final surgery," the point in the film where she really expresses her transgender identity is when she sings at the lesbian nightclub. Thomas (2001) suggests that although "any physical space can function as a dramaturgical stage, many films use theatres, actors, and performances within their narrative worlds to give greater prominence to such con-cerns and to stand as a metaphor for the wider narrative world" (41). This is certainly the case with *Better Than Chocolate*. In most of the scenes depicting onstage performance, the characters in this film seem to be expressing something about themselves as performers. This is no-where more true than when Judy sings.

The song that Judy performs, "I'm Not a Fucking Drag Queen," was written by an actual transgendered woman in Vancouver who performs the song live herself, according to Wheeler (commentary). As she sings, Judy walks around the club and interacts with women in the audience, who seem receptive to her attention (although Frances does appear to be flustered). Everyone in the club applauds as she concludes her song.

In this dramatic scene Judy is expressing a facet of identity that could be misinterpreted by others (as evidenced by some reviewers referring to her as a drag queen). Thomas (2001) suggests that characters onstage can be seen as "revealing something like an 'authentic' self that lies beneath the surface presentation" (40). Judy is asserting with this scene that her "authentic" self is transgendered, even if her surface presentation could be read as that of a drag queen. The film, and the lesbian nightclub within the story world, are giving her space to assert herself as she would like to be seen and understood by others. The thunderous applause in the club clearly indicates that the people in this filmic lesbian community accept and appreciate her identity. This acceptance clashes with real-life instances of rejection of trans women in lesbian communities, and with the woman in the story world who actually does hit Judy. The result is a narrative that privileges the position of accepting trans women in lesbian communities, with a resultant defiance of real life dominant attitudes and a marginalization of those who oppose such acceptance.

'SHE' AND 'HER' AND 'SISTER'

In addition to the support she gets from the audience at the club, most characters in the film state onscreen at some point how they feel about Judy. Maggie and Kim, the film's central lesbians, welcome Judy into their community without regard to her transgender status. When another woman says that "she's not a woman," Maggie replies, "she *is* a woman, and she's our friend," and Kim adds, "yes, she is" when the woman disagrees. Besides this scene, where their vocalization of Judy's womanhood is emphatic, they seem to just accept Judy's trans-ness as just another part of her.

Not everyone in the onscreen lesbian community accepts Judy unconditionally, however. For example, Judy's love interest, Frances, is less than receptive to Judy's advances. The nature of her rejection brings up one of the issues surrounding trans inclusion in the lesbian community: are you still a lesbian if you are dating a trans person? The issue's more public face is that of lesbians whose partners choose to transition from female to male. The partner who is not transitioning has to think about what it means for her identity that her partner identifies as male–can she be a lesbian and still love a man? If she chooses to retain her lesbian identity, the transitioning partner needs to think about what it means to him that his partner thinks of herself as a woman who loves women, and what it means to his identity as a man. *Better Than Chocolate*, however, explores a different issue–that of a nontrans lesbian who becomes romantically involved with a trans lesbian. This is only a problem for a woman who sees trans women as not "real" woman, and therefore unsuitable to date.

Frances begins the film by rejecting Judy's romantic advances, if not Judy herself. When Judy is talking about moving into a new condo, and asks for advice, Frances calls the complex "cheap and tawdry goods, masquerading as quality." The rhetoric surrounding those in lesbian communities who oppose trans inclusion often focuses on the idea that trans women are mimicking or masquerading as *real* women, or that they are inferior copies of real women. Again, this idea requires the belief that there is an *original* version of femininity that trans women copy. This concept requires a rejection of Butler's (1990) claim that "there is no gender identity behind the expressions of gender: that identity is performatively constituted by the very 'expressions' that are said to be its results" (25). Considering the tendency for some people to use the language of original/copy, it seems likely that Frances's remark is a passive-aggressive insult directed at Judy. Through an offhand remark, she is able to express her disapproval of Judy without being direct. It is particularly probable that Frances meant it as an insult, and that Judy understood this, because Judy immediately replies to the comment by saying "comme moi."[4]

Later in the film, after Frances invites Judy back to her apartment, she shies away from Judy's advances, saying that she "need[s] a little more time." Wheeler said that she wanted it to be clear that Frances "rejects him [*sic*], but you have to feel that it's more to do with her than with him" (commentary). Specifically, that Frances is rejecting Judy not because of who Judy is but because of internal struggles Frances is dealing with, struggles that likely involve resolving her lesbian identity with Judy's transition.

The line is deliberately ambiguous so that Frances's rejection can be read a number of ways, depending on what you as a viewer want to see. It could mean, for example, that Frances needs more time to be ready for a relationship, that she needs more time to adjust to having a trans lover, or that she needs more time for Judy to change. She may not think of Judy as a *real* woman, and is dealing with what that implies about herself if she decides to get involved with Judy. She verbalized a belief that Judy isn't really a woman earlier when she insinuated that Judy was an inadequate replica. While she seems to be coming around to her attraction to Judy, she could still be wrestling with these issues. She clearly doesn't want to hurt or alienate Judy by her behavior, in any case, because she nervously tells her that she had a really good time and agrees to call her.

The contrast between Judy and Frances is striking. Judy uses sweeping hand gestures and open body language, but Frances makes choppy, nervous gestures and seems closed-off physically. Where Judy wears elegant, flowing garments, Frances wears boxy, confining clothing. According to Lurie (2000), "bundled-up or buttoned-up clothes (if not figure-revealing) are felt to contain a tight, erotically held-in person" (231). This is further illustrated when Frances says that she "has never been what you would call sexually adventurous," and that she's "never had a threesome . . . I've barely had twosomes. I have had three girlfriends, all of whom were exactly like me." Her restraint further complicates her reaction to Judy's advances, as she seems like someone who would be nervous and edgy when anyone pursued her romantically, even someone whose gender identity Frances did not need to internally reconcile.

Besides Frances (and to a lesser degree, Lila), there is only one woman in *Better Than Chocolate* who rejects Judy. This character (the end credits call her Woman in Washroom) approaches Judy in the bathroom of the nightclub and forcibly tries to get her to leave.

About this scene, Wheeler says:

> This scene is one that's really upset a lot of people, in fact some people don't want it in the movie, but we felt really strongly that we wanted to show all the sides of the community. It wasn't one big happy loving family and there are divisions amongst the lesbian population. But also just the trials and tribulations that somebody like Judy would live through. We wanted I guess to be on a serious level with that issue and all the transgendered women that we talked to said that acceptance by other women was one of the hardest things about being transgendered. (commentary)

She also states that she "talked to many transgendered women and based the incident on a real life episode and one that was not the worst by far" and that the scene was "one of the heavier moments which hits people as shocking because of its contrast to the humour of the rest of the film" (qtd. in Gellman, "Venus," para. 10).

I WON'T LET YOU FORGET IT

Through Kim and Maggie's open acceptance of Judy, Frances's coming around from rejection to love, and the obviously negative portrayal of Woman in Washroom, it's clear that the film makes a strong statement about how transgendered women should be welcomed and accepted into lesbian communities. Everyone loves and accepts Judy, and those who don't either come around or get put into their place.

A closer look at the movie, however, reveals that a statement is being made not only about including trans women into lesbian communities but about those who, for whatever reason, have reservations about doing so. The film makes an unequivocal statement about trans inclusion–ultimately, anyone who voices resistance either changes her stance or is put in her place. In addition, resistance is coded negatively in every situation. Frances is thought of as ridiculously old-fashioned. She sees the error of her ways, however, and eventually comes around. Woman in Washroom is one-dimensional, and seems to be in the movie only to play the part of an unsympathetic, violent transphobe. There are no characters who voice affection and support for Judy, but would prefer she not participate in particular events.

This is particularly interesting since "womyn's festivals" are mentioned twice in the film. Frances runs into a woman she hasn't seen since the "Womyn's Music Festival," and at the end it is revealed that Kim and Maggie "hit every womyn's festival in North America." One of the most hotly contested sites in the world of trans inclusion and women's spaces is women's festivals, most notably the Michigan Womyn's Music Festival. Morris (1999) writes about "today's hot issue: transgendered festiegoers" (171), and suggests that one of the main arguments against trans-inclusion is that the "celebration of female life and energy that is festival culture seems mocked by the inclusion of men who have *selected* female identity" (175, emphasis original). Emi Koyama (2002), on the other hand, suggests that when "an action or a pattern of actions negatively impacts a particular community's rights disproportionately, it is considered discriminatory regardless of the intention" (10), and that there are no logically sound arguments against trans-inclusion in women's spaces.

Judy doesn't object to anyone's attendance of these festivals onscreen, and Kim and Maggie attend them instead of boycotting them. The film never indicates that Judy would not be welcome at these festivals. Bringing this issue up would be a way to bridge the filmic world to the real world and suggest ways in which transpeople are excluded from lesbian communities. This omission seems like a way to try to appease more members of a lesbian audience: supporting trans women's right to participate in lesbian communities without condemning festival spaces. It's a thin line to walk.

This easy simplification of Judy's experience in her community is explained, however, by Dyer's (2002) concept of the film utopia. "The capacity of entertainment to present either complex or unpleasant feelings," he suggests, is such that it "makes them seem uncomplicated, direct and vivid, not 'qualified' or 'ambiguous' as day-to-day life makes them" (25). That is, Judy doesn't fight with Kim and Maggie about attending womyn's festivals, and they in turn accept her unconditionally as one of their own, without the nuances or complications inherent in such situations in real life.

TENDER TRANSGENDER HEART

While the film sends a strong message of trans inclusion, its portrayal of a transgendered lesbian is not entirely positive. Judy, for example, seems to need help and validation from the "real" lesbians in her community. When Woman in Washroom is attacking her, she asserts her right to be there but allows her to hit her with her purse. Maggie and Kim have to come to her rescue. Wheeler says that they "wanted her to be a person who doesn't necessarily stick up for herself," but who later sticks up for others.

In addition, Frances's change of heart regarding Judy's romantic pursuit could be read not as a realization that Judy is really a special woman deserving of her love but as settling for a transwoman since she seems unable to attract "real" lesbians. Besides the fact that Frances seems unaccustomed to requests for dates, it is revealed that an ex of hers, Bernice, left her for another woman. She admits to Judy that she hasn't had much success with women in her lifetime. There's nothing in the narrative to prevent the reading that she's making do with Judy, although it does not actively assert or state this possibility.

The point is, the film tries to incorporate and please many different approaches to lesbian life, but ultimately fails to portray any one to satisfaction. The story world is mainly pro-trans, but the woman they're supporting is sweet and good-natured. One wonders if everyone would be as quick to accept an MTF who was as angry or abrasive as Woman in Washroom or who was less feminine than Judy. In the interest of creating a pro-trans lesbian community, the filmmakers didn't allow any nuances of reservation. In this fictional Vancouver, you either accept and welcome transwomen into your community or you're an angry, violent person who doesn't even merit a first name.

NOT SOME MIDNIGHT RACKET

Better Than Chocolate was incredibly successful, both commercially and critically. It continues to show up on favorites lists on lesbian-themed Websites, where women swap tips on how to reenact the body-painting scene with their lovers. In a world where the issue of trans inclusion is very much active in lesbian circles, it seems that any message in a film this popular would have an impact on viewers' opinions. After all, who wouldn't rather side with attractive, happy Kim or Maggie than with unhappy, nervous Frances or the furious Washroom woman?

Wilton (1995) suggests that the "lesbian 'community' is . . . a complex and conflict-ridden social context for the production and consumption of the moving image" (14), and that "there are important social, cultural, and political subjectivities that attach to being lesbian, and that impact upon the subjectivity of 'real' lesbians who watch films/videos" (5). For these reasons, lesbians can celebrate "a romantic comedy where it's no big deal that a transgendered . . . woman named Judy falls in love with a celibate lesbian" (Connors, 1999, "Sweet Lesbian Tale," para. 9), because it offers us a hopeful, loving view of lesbian communities. I would, however, voice a concern that this view comes at the expense of both women who love and accept transwomen, but have reservations about them in every conceivable lesbian space, and of transwomen who themselves can't or don't care to perform Judy's level of femininity. While the film broke ground with its portrayal of a lesbian community enthusiastic about accepting a trans woman, someday I suspect it will be remembered as a quaint, superficial but entertaining entry in the history of lesbian-centered film.

ISSUES AND QUESTIONS FOR FURTHER CONSIDERATION

1. To what extent should films reflect the social climates of the cultures in which they are created and seen? Should *Better Than Chocolate* have considered more of the challenges trans women face in lesbian communities?
2. How positive of a representation of trans women is Judy?

3. How would you compare the pairing of Judy and Frances to Kim and Maggie in the film? Does this reveal a bias on the part of the production staff?
4. Most of the characters in *Better Than Chocolate* are white. Does this impact the way the film deals with the issue of including trans women in lesbian communities? Would it be different if the community were multiracial, Black, Asian, or Latino/a?
5. How does the film address issues of gender and sexuality as both identity-based and performative? How do the characters convey their gendered and/or sexual identities? How does this reflect the world you live in?

NOTES

1. In this piece, I use the prefix "trans" as an inclusive term to refer to transgender, transsexual, and genderqueer, which all describe persons trying to find space in lesbian communities, and for whom lesbians are trying to find space within their communities.
2. My section titles come from songs different characters perform onstage in the film.
3. Transsexuals are not the only group Lurie generalizes about; gay men, lesbians, and feminists also get a broad gloss.
4. "Like me"

REFERENCES

Butler, J. (1990). *Gender Trouble*. London and New York: Routledge.
Connors, J. (1999, September). 'Chocolate' Fudges Parts of Sweet Lesbian Tale [Review of the motion picture *Better Than Chocolate*]. *The Plain Dealer* [online]. Retrieved June 5, 2004. Available: http://www.academic.marist.edu/tmurray/CCOX/newsbtc9.htm
Dyer, R. (2002). *Only Entertainment* (2nd ed.). London and New York: Routledge.
Gaines, J. M. (2000). Dream/Factory. In C. Gledhill & L. Williams (Eds.), *Reinventing Film Studies* (100-113). London: Arnold.
Gellman, D. (1999, August 12). Venus of Mars Bars [Review of the motion picture *Better Than Chocolate*]. *Xtra!* [On-line]. Retrieved June 5, 2004. Available: http://www.xtra.ca/site/toronto2/arch/body251.shtm
Koyama, Emi. (2002). *A Handbook on Discussing the Michigan Womyn's Music Festival for Trans Activists and Allies*. Portland: Confluere Publications.
Lurie, A. (2000). *The Language of Clothes* (2nd ed). New York: Henry Holt and Company.
Morris, B. (1999). *Eden Built by Eves: The Culture of Women's Music Festivals*. Los Angeles: Alyson Books.
Stacey, J. (1995). 'If You Don't Play, You Can't Win': *Desert Hearts* and the Lesbian Romance Film. In T. Wilton (Ed.), *Immortal Invisible: Lesbians and the Moving Image* (92-114). London and New York: Routledge.

Thomas, D. (2001). *Reading Hollywood: Spaces and Meanings in American Film.* London and New York: Wallflower.

Warwick, A. & Cavallaro, D. (1997). *Fashioning the Frame: Reframing the Boundaries of Sex.* New Brunswick, New Jersey: Rutgers University Press.

Wheeler, A. (Director), Thompson, P. (Writer), & McGowan, S. (Producer). (1999). *Better Than Chocolate* [Motion picture]. Canada: Trimark Pictures.

Wilton, T. (1995). Introduction: On Invisibility and Mortality. In T. Wilton (Ed.), *Immortal Invisible: Lesbians and the Moving Image* (1-19). London and New York: Routledge.

Debating Trans Inclusion
in the Feminist Movement:
A Trans-Positive Analysis

Eli R. Green

SUMMARY. The debate over whether or not to allow, accept, and embrace transpeople as a segment of the feminist movement has been a tumultuous one that remains unresolved. Prominent authors have argued both sides of the dispute. This article analyzes the anti-inclusion feminist viewpoint and offers a trans-positive perspective for moving toward a potential resolution of the debate. *[Article copies available for a fee from The Haworth Document Delivery Service: 1-800-HAWORTH. E-mail address: <docdelivery@haworthpress.com> Website: <http://www.HaworthPress.com> © 2006 by The Haworth Press, Inc. All rights reserved.]*

KEYWORDS. Transfeminism, feminism, Janice Raymond, lesbian separatism, lesbian communities, identity of oppression, identity politics

---

Eli R. Green is a gender warrior and transactivist with a master's degree from Claremont Graduate University in Applied Women's Studies. Eli is webmaster for the popular website Trans-Academics.org and serves on the board of the National Transgender Advocacy Coalition.

Address correspondence to: Eli R. Green, 151 First Avenue, #228, New York, NY 10003 (E-mail: eli@trans-academics.org).

[Haworth co-indexing entry note]: "Debating Trans Inclusion in the Feminist Movement: A Trans-Positive Analysis." Green, Eli R. Co-published simultaneously in *Journal of Lesbian Studies* (Harrington Park Press, an imprint of The Haworth Press, Inc.) Vol. 10, No. 1/2, 2006, pp. 231-248; and: *Challenging Lesbian Norms: Intersex, Transgender, Intersectional, and Queer Perspectives* (ed: Angela Pattatucci Aragon) Harrington Park Press, an imprint of The Haworth Press, Inc., 2006, pp. 231-248. Single or multiple copies of this article are available for a fee from The Haworth Document Delivery Service [1-800-HAWORTH, 9:00 a.m. - 5:00 p.m. (EST). E-mail address: docdelivery@haworthpress.com].

Available online at http://www.haworthpress.com/web/JLS
© 2006 by The Haworth Press, Inc. All rights reserved.
doi:10.1300/J155v10n01_12

We as feminists owe it to ourselves . . . to deconstruct and oppose . . . trans politics. In a feminist analysis they are, to put it simply, on the wrong side. In opposition to feminism.

–Charlotte Cronson, "Sex, Lies and Feminism"

Nothing upsets the underpinnings of feminist fundamentalism more than the existence of transsexuals. A being with male chromosomes, a female appearance, a feminist consciousness, and a lesbian identity explodes all of their assumptions about the villainy of men. And someone with female chromosomes who lives as a man strikes at the heart of the notion that all women are sisters, potential feminists, natural allies against the aforementioned villainy.

–Patrick Califia, *Sex Changes: Transgender Politics*

As the quotes above illustrate, feminism has historically been and is currently still divided on the issue of whether or not to accept transpeople (particularly transwomen) and include relevant trans issues as a part of the feminist movement.[1] As trans identities have become more visible and prominent in society, the tensions between feminists and transpeople have also escalated. This article analyzes the underpinnings of '*anti-inclusion feminism*' (the feminist politic of purposely and actively excluding transpeople and trans issues from feminist action, ideology and space) from a transfeminist perspective.[2] It maintains that the anti-inclusion feminism is primarily motivated by ignorance and misinformation about trans identities, transpeople and trans community/culture, as well as being further triggered by a fear that trans-inclusion could potentially undermine feminist theory and ideology.

I accomplish this by first briefly examining the historical moment in which this debate came to the forefront of the feminist movement, as well as offering an analysis of anti-inclusion feminist theory put forth at the time by Janice Raymond in *The Transsexual Empire: The Making of the She-Male*. Moving forward to present day, I show how Raymond's problematic and transphobic anti-inclusion theories are still cornerstones of this conflict through an examination of the mission statement of QuestioningTransgender.org–a new Website dedicated to opposing "the hegemony of transgender politics among lesbian and feminist communities."[3] I scrutinize the arguments brought forth by Questioning

Transgender.org from a trans-positive perspective and contrast them with various trans narratives. I then offer the model of activism created and operated by the Gender Political Advocacy Coalition (GenderPAC) as an alternative to current anti-inclusion feminist politics. Finally, I seek to deconstruct the *'feminist identity of oppression,'* through an assessment of the ideologies on which feminism is based, with the hope of creating space for future discussion about the part that feminism plays in the oppression of transpeople.

THE HISTORICAL ROOTS OF SEPARATION

As feminist theory matured and became an integral part of the activist-based feminist movement of the seventies, it became heavily entrenched in a period of intense political correctness, based largely on the then recent focus of having a theoretical base to feminist action (Butler, 2004). This warranted a phase of extreme internal political change in the movement, which was manifested outwardly in part by the androgynous-feminist look of the day. Traditional feminine attire was seen, both literally and symbolically, as oppressive to women and was thus abandoned in favor of androgynous dress as a part of a greater political statement (Meyerowitz, 2002).

This was of particular challenge to feminine women (trans and cisgender) who enjoyed and embraced feminine self-expression as an integral part of their identity.[4] The paradigm shift in "proper" feminist attire brought feminine transwomen to the visual forefront of the feminist movement. Feminine transwomen were singled out as traitors to the feminist movement for their perceived rejection of androgynous dress and the underlying feminist ideology (e.g., rejection of traditional or oppressive femininity).[5] In the search for visible political distance from feminine transwomen and transwomen as whole, some non-trans feminists worked to create a fissure between the two types of "woman" by highlighting the perceived illegitimacy of labeling transwomen as "women," based on the assignment of male at birth. As this fissure was established, transwomen's feminist identities became suspect, diminishing the space for transwomen as a part of the feminist movement. This metaphoric attitude transitioned to the physical realm through the practice of actively prohibiting transwomen from feminist spaces.

While there were other significant factors that played a part in this ostracization, the shift in feminist ideology and dress served as a visible line of demarcation for the volatile beginning and led to the attempted eradication of trans identities from the feminist movement. It served as a major incendiary component in the now more than quarter-century debate over the validity of trans identities, politics, and transpeople's rights to access feminist space.

RAYMOND'S PROBLEMATIC LEGACY

The desire to expunge trans identities from the feminist movement and feminist space is undoubtedly influenced by the infamous legacy of anti-inclusion transphobic-feminist Janice Raymond. In 1979, Raymond published *The Transsexual Empire; The Making of the She-Male,* bringing tangible voice to a decade of vocal dissent and opposition of trans inclusion. While in many circles the book is now considered to be a pseudo-academic piece due to its circular logic and questionable research methods, it nevertheless stands as a cornerstone of anti-inclusion feminist politics.[6] When the book was published, it was the first prominent piece to offer a perspective on why gender-variant persons should be excluded from the feminist movement in a time where counter-arguments were not readily accessible and were often dismissed as the words of an unreliable "enemy" (Meyerowitz, 2004). Thus, *The Transsexual Empire* became the largely unchallenged leading voice of the feminist movement's ideology opposing trans-inclusion.

Despite being published over twenty-five years ago, the book is still highly relevant to the current discussion of trans-inclusion. Many of the ideas forwarded by Raymond in 1979, such as what defines "woman," the medicalization of gender and sex, the social and biological legitimacy of trans identities, the placement of biological influences in a social constructionist feminist movement, and the purpose and sanctity of feminist space, are still cornerstone questions of today's trans-inclusion debate.

One of the major faults of *The Transsexual Empire* rests in the research methods used to support the assertions in the book. The book is based in large part on medical publications presenting transsexuals as persons who are "born into the wrong body." At the time of *The Transsexual Empire's* publication, this represented the sole model of medically acceptable transsexuality, a framework that refused to accept that trans identities have a base in anything other than patriarchal motives of oppression. According to Raymond's perspective, any person seeking medical "treatment" for a gender-variant identity is enforcing the rigidity of the gender binary by seeking to become the opposite gender.

As Sandy Stone points out in her rebuttal to Raymond's work, *The "Empire" Strikes Back: A Post Transsexual Manifesto*, the creation of the archetypal transsexual was a direct result of the medical establishment itself. In order to meet the requirements of the highly prominent Harry Benjamin Standards of Care (HBSOC), the dominant set of guidelines used by clinicians for assessment and management of gender-variant clients, a patient must fulfill certain criteria of transsexuality in order to receive hormone therapy and/or surgery.[7] HBSOC operates on a highly medicalized gateway model that gives clinicians complete power in deciding who can and will receive hormones and/or surgery. Under the HBSOC model transfolk have been refused services for a perceived inability to pass, having a non-heterosexual orientation, or in some instances, being married (Meyerowitz, 2002). Patients who can even afford to be seen by the select doctors who have an understanding of and willingness to work with gender variance have little voice in the process and little to no recourse if they are not seen as "appropriate." Due to the prominence of the HBSOC standards, many patients alter their personal narratives out of necessity in order to obtain surgical or hormonal services from medical practitioners. This creates a significant, and unaddressed, gap between Raymond's transphobic theories and the realities of trans existence and identity.

The self-selective nature of these medical biographies has essentially painted a picture that all gender-variant folk have a sense of "being born into the wrong body" and seek gender clarity by becoming an unambiguously recognized member of the "opposite sex." This description is one of a select portion of the trans community, and does not include the wide variety of persons with gender-variant identities who do not fit this archetype, such as those who identify as genderqueer, two-spirited, intergender, pangendered, and, in some instances, transgender.[8] These people can and do seek medical services for different reasons than to "become the opposite sex." This wider perspective doesn't even account for gender diverse people who do not identify with any of these or other trans labels and do not seek any medical services related to identity-confirmation. This wide area of perspectives and identities is not included in the current predominant medical model and is not accounted for in Raymond's methods or text.

In the *Transsexual Empire*, Raymond asserts that transpeople, and transwomen in particular, lack a feminist consciousness of patriarchal gender roles and are suffering from a patriarchal, medically invented, psychological illness (Raymond, 1979). She blames the patriarchal medical community for the creation of gender-variant identities. However, the medicalization of gender is a relatively recent phenomenon and cannot account for the historical presence of gender-variant identities. Research indicates the existence of individuals that would meet the current standard for gender-variant or trans identities dating back as far as the Old Testament (Feinberg, 1996). These individuals existed in a time when the patriarchal value-system of modern medicine was not a factor, thus creating a serious flaw in Raymond's argument of causality.

In *The Empire,* Raymond blames transpeople for enforcing the gender binary by transitioning from one gender to another. In doing this, she does not consider that it is not necessarily the *choice* of gender-variant people to enforce a gender binary. Instead, it is the medical community that enforces the duality of gender through definitive sex assignment, intersex sexual assignment surgeries,[9] and reliance on the Harry Benjamin Standards of Care to judge "true transsexuality." Gender-variant folks have nearly a complete lack of viable alternative options outside of submitting to the authority of the medical establishment. Through her reliance on the medical field for her perspective on gender variance, Raymond views transsexuals as merely buying into the inflexibility of gender and completely misses the reality of a gender-variant person's existence.

Another major factor to be considered is that Raymond built her book on the fallacy of denying the legitimacy of socially constructed transwomen's identities while at the same time affirming socially constructed non-trans women's identities. According to Raymond, biological women are the only people who get to define what 'woman' is, or is not:

> We know who we are. We know that we are women born with female chromosomes [sic] and anatomy. . . . (Raymond, 114)

Here Raymond uses chromosomal sex and thus a medical assessment of gender to enforce her position that "woman" (and thus the right to be a woman) is an identity bestowed by biology and cannot legitimately be *chosen.* She fails to acknowledge or realize that her definition of woman is derived from the very same sex and gender binaries that she faults and are in large part enforced, if not created, by the same medical establishment which she criticizes for "creating" trans identities. It is therefore illogical to oppose the construction of trans identities, when according to her social constructionist views of gender, the identity of "woman" is created in the same way. She attempts to use medical legitimacy of 'woman' to draw a line in the sand to create an infallible reason for exclusion of transpeople, specifically transwomen. However because of the contradictions in her logic, she is no more successful or accurate than patriarchal practices of oppression. Additionally, in the quoted text, Raymond inherently assumes that "woman-ness" is necessarily derived from being born female. While this assumption that gender is contingent on sex is hegemonically axiomatic, deeper analysis of this presumed link calls into question the idea that gender expression and identity is completely driven by biological sex. Judith Butler (2004) addresses this in *Undoing Gender*.

> [T]he critique of male-to-female transsexuality has been centered on appropriation of femininity, as if it belongs properly to a given sex, as if sex is discretely given, as if gender identity could and should be derived unequivocally from presumed anatomy. (pg. 9)

Butler not only questions the base of gender, but also the way in which 'sex' is assigned and is used as validation for the eradication of trans identities from feminist space. At this point it is impossible to say what, if any, connections there are between biological sex and gender identity due to a lack of understanding of how gender and sex intertwine. Eliminating this assumption is a key step in gaining a complete and accurate understanding of how gender identities are developed.

While Raymond's text focuses almost completely on the illegitimacy of transwomen, and is thus the central focus of analysis here, it is also important to note her reasons for doing so. In the introduction to the 1994 version of *The Empire* Raymond states:

> Transsexualism remains, as in 1979, a largely male [read: transwomen] phenomenon. Female-to-constructed-males [sic] are relatively rare. For example of the transsexual surgeries that are performed at the University of Minnesota's Program in Human Sexuality . . . 85% are male to female. More interesting are the reasons why. . . . Women [read: transmen] have had a political outlet, that is, feminism. (pg. xiii)

Raymond suggests transmen do not exist because there are not as many transmen accessing transsexual surgeries. This is interesting, given that she acknowledges that surgeries available to transmen are not as successful or accessible as those available to transwomen. However, the data she cites does not take into account transitioning surgeries (e.g., top or chest) and hormone treatments that may not occur as a part of a trans-

sexual medical care program. Since many transmen develop male sec-
ondary sexual characteristics under varying amounts of time on
testosterone, the surgical statistics do not give a clear picture of the frequency of transmen's existence.

Raymond's hypothesis that feminist consciousness prevents or in-
hibits the existence of transmen is equally curious. She seems to allege
that female-bodied feminists will not identify as transmen *because they
are feminists*. The converse, that transmen cannot be feminists because
they are trans, would also follow from this premise. However, Raymond's inference is problematic on several levels. Primary among these
is that it places feminism in diametric opposition to trans existence,
while ignoring the fact that feminist consciousness is defined by intellect and reason, not by biology. Her statement assumes that transmen
who transition from female to a masculine-male gender expression are
doing so without being informed of feminist consciousness. As Califia
(2003) points out in his analysis of *The Empire*, Raymond maintains
that transmen are transitioning as a means of escaping the oppression
faced by women, which allegedly would be eradicated by access to feminism. Judith Butler comments on the fallaciousness of this escaped oppression theory.

> The view that transsexuals seek to escape the social condition of
> femininity because that condition is considered debased or lacks
> privileges accorded to men assumes that female-to-male (FTM)
> transsexuality can be definitively explained through recourse to
> that one framework for understanding femininity and masculinity.
> It tends to forget the risks of discrimination, loss of employment,
> public harassment, and violence are heightened for those who live
> openly as transgendered persons. The view that the desire to be-
> come a man or transman or to live transgendered is motivated by a
> repudiation of femininity presumes that every person born with fe-
> male anatomy is therefore in possession of a proper femininity
> (whether innate, symbolically assumed, or socially assigned), one
> that can either be owned or disowned, appropriated or expropri-
> ated. (Butler, 2004, pg. 9)

Butler illuminates the disparity between perceived and actual trans existence and expression in Raymond's work. A growing body of work documents individual struggles to reconcile this internalized feminist
doctrine with person transitions from female-to-masculine gendered
expressions.[10] Jamison Green verbalizes the challenges of maintaining

such a precarious balance during his own transition in *Becoming a Visible Man*.

> I had to understand my part in that system of inequity, whether I occupy a female place and a masculine role or a male place and a masculine role. I needed to understand what it would–really *mean*– to change places: what responsibility would I have for maintaining or deconstructing traditional gender roles once I transitioned? (2004, pg. 23)

Another example of this consciousness and conflict is voiced by a young transman and feminist in an article appearing in the *San Francisco Chronicle*.

> Kaisaris, as a feminist, says the entry into the society of men makes him somewhat uncomfortable. Though he is now afforded certain male privileges, he finds himself in the quirky position of becoming a man-hating man. "It's like being inducted in an underground society," he says with due seriousness. "My responsibility is to become a decent man." (Rafkin, 2003)

Kaisaris not only shows that it is possible to have a transmasculine identity while maintaining a feminist consciousness, but also emphasizes a responsibility to not become an oppressor of women. The two previous quotes are representative of several that contest Raymond's assertion that it is not possible to be both trans and feminist, as well as document the challenges of existing in these multiple, and sometimes conflicting, identity spaces.

CONTINUING THE LEGACY

In her work, Raymond presents the idea that transwomen are deviant men on a mission to destroy or at least usurp the success of the feminist movement. Because of this clear anti-feminist mission, see calls on all feminists to eradicate any gender-variant presence in the movement as a means of protection (Raymond, 1979).[11] This is a very clear anti-inclusion sentiment that is currently being echoed by several other anti-inclusion feminists. One such example is the Website Questioning Transgender.org, which is dedicated to the anti-inclusion perspective. Their mission statement reads as follows.

We stand opposed to trans-politics [Read: politics of trans-inclusion] because:

- [These politics] include the insistence that transgender and transsexual individuals be served by organizations designed by and for women without regard for the concerns, desires, and interests of the women involved.
- [These politics] undermine our ability to understand that the gender classes of men and women are socially created.
- [These politics] deny or ignore the social, economic, and power differentials between these two classes that amount to the oppression and domination of one over the other.
- [These politics] fail to address the significant problems of male power and male violence across the world, including violence against women as well as violence against transgender people. (From: http://questioningtransgender.org)

The site's mission statement inaccurately presumes that there is only one trans-inclusion perspective, or trans-politic, to critique. To label all transpeople and their ideas under the assumption that there is only one "trans-politic" is vague, and suggests a lack of knowledge about the diversity of trans culture and politics. Further, the site offers neither research-based nor factual support for the premises in the mission statement, weakening all the arguments contained within the Website.

One of the overriding themes of the mission statement is the criticism that trans-activism is not operating under a complete and total feminist consciousness, or that trans-activists are not feminist enough. The latter criticism is ironic considering that these are the same people arguing against trans-inclusion in a feminist movement, an inclusion that would theoretically encourage or require such a consciousness. The idea that transpeople do not spend enough time evaluating issues against women is a recurring theme, one that does not take into account that anti-inclusion feminists spend more time dismissing rather than evaluating trans issues potentially related to the feminist movement. The third tenet of the mission statement reads: "[Politics of Trans-Inclusion] deny or ignore the social, economic, and power differentials between these two classes that amount to the oppression and domination of one over the other." This demand of engaging feminist analysis results in a continuing and inappropriate burden shifting onto transpeople. It places undue expectations on transpeople to single handedly overcome massive barriers, so that they might be able to engage in feminist consciousness and analysis of their very existence.

QuestioningTransgender.org's statements suggest that minority groups facing the severest oppression should be responsible for analyzing (and eradicating) their own oppression, rather than placing the necessary analysis of oppression on those who engage in or enforce the oppression of others. That is not to say that minority groups, particularly transpeople, who face substantial oppression should not be cognizant of their part in the oppression of others and work to end said oppressions. However, it should not be the *one-sided* responsibility of transpeople to take on cisgendered women's oppression as the central core of its own work. This is especially true when a more privileged minority group (in this case anti-inclusion feminists) is actively engaged in the oppression via the exclusion of others–specifically transpeople.

The second point of QuestioningTransgender.org's mission statement reads: "[Politics of Trans-Inclusion and Transpeople] undermine our ability to understand that the gender classes of men and women are socially created." This is a curious statement, as it literally blames transpeople for undermining the stability of socially constructed gender. One of the main arguments of the Website (and of Raymond) is that transpeople are problematic because their identities and gender expressions are *confirming* the rigidity of the gender binary. However, the QuestioningTransgender.org Website does a 180 degree turn-around and censures transpeople for blurring the lines of gender. Anti-inclusion feminists are creating a no-win situation by condemning transpeople for both confirming and blurring the gender binary.

This section of the mission statement inadvertently highlights a potential undercurrent of anti-inclusion feminist discomfort over trans identities. If science advances to uncover a biological contribution to gender variance, this could undermine the assertion that gender as a class is completely socially constructed. The demonstration of biological contributions could drive a movement to reevaluate feminist majority theory, which is based in large part on strong social-constructionist views of gender. Logically, this suggests that one of the main motivations behind anti-inclusion feminism is fear, especially a fear that by their mere existence, transpeople could and do call into question the very foundation of the feminist movement.

FEMINIST IDENTITY OF OPPRESSION

Like that of most activist movements, feminism is based on an *identity of oppression*. The movement cannot exist without the oppression it seeks to end. While in and of itself this identity is not a negative one, anti-inclusion feminists use an identity of oppression as a shield from trans inclusion and the aforementioned possible undermining of feminist foundations. *We are oppressed, we are the only oppressed gender because gender variance is not a valid gender expression, and our oppression takes precedence over all others.* Through this, anti-inclusion feminists also fall into the greater societal pattern of refusing to acknowledge one's own role in the oppression of others.

It is very easy for anti-inclusion feminists to acknowledge that transpeople are oppressed. However, this acknowledgement does not recognize that because women, feminists, and anti-inclusion feminists have more privilege than transpeople, that they are oppressors. That is not to say that on an individual level one cannot challenge and work against these systems of oppression, although as a part of a more privileged social class, it is impossible to leave that privilege completely behind. Nowhere on the Website QuestioningTransgender.org, or in any other anti-inclusion text reviewed for this article, was there an examination of the role that feminists, in particular anti-inclusion feminists, might play in the oppression of gender-variant people. This omission is one concrete example of how anti-inclusion feminists have adopted the identity of oppression to deflect constructive criticism and introspection regarding their relationship with and against gender-variant identities.

This has not always been the ethos of the feminist movement. An example is the shift in lesbian inclusion in the feminist movement over the course of the 1970s. In the early part of the decade, many feminists felt that including lesbian issues in the feminist movement would ultimately hinder its success. At the time, lesbians were considered to have distinct issues that would draw attention away from the issues of the heterosexual majority. Out lesbians who refused to congeal with the ideals and actions of the feminist movement were outcast for their differences (i.e., sexual orientation) rather than embraced for their similarities (i.e., gender).

In opposition, lesbian communities accused feminists of engaging in heterosexist and patriarchal lesbian baiting. Extreme persistence and consciousness-raising by lesbians resulted in a diametric reversal by the end of the decade. Lesbianism was seen as the ultimate feminist representation for its perceived complete abandonment of male presence and dependency. This consciousness raising was so successful that it became en vogue for heterosexual women to forgo sexual contact with men and proclaim themselves "political lesbians," in order to further the complete eradication of patriarchal oppression of women. Lesbian voices were heard and respected in no small part because the category of "lesbian" inherently included "woman." While including lesbian concerns on a feminist platform may have originally been problematic, ultimately it did not complicate the movement's centeredness around "woman" as the only oppressed gender.

Consciousness-raising about trans identities and inclusion within the feminist movement has been significantly more complicated. In addition to the fear that biologically based transgender identities challenge significant amounts of feminist theory, trans-inclusion would require acknowledgement that "woman" is not the only oppressed gender. It would force feminists to recognize gender-variant persons as *validly gendered*. Cisgendered women have the distinct privilege of being a part of a legitimate social class–woman. While the class of woman is certainly one of a patriarchally oppressed "other," the legitimacy of its right to exist is not routinely under attack. This is a privilege that, as many documented cases of violence against transpeople have shown, can mean the literal difference between life and death.[12]

PROVING THE POSSIBLE:
THE GenderPAC MODEL

The Transsexual Empire by Janice Raymond and Questioning Transgender.org's mission statement highlight a related and recurring idea that trans-activism seeks to take over the feminist movement and replace its doctrine with gender-variant centric activism. However, those trans-activists that are even interested in pursing feminist collaboration seek to work towards the eradication of *gender-based oppression*. This approach concentrates on the end of oppression to women *and* transpeople. It does not suggest that the entire focus of feminism become trans-centric; rather, it requires the dedicated analysis of where these movements might successfully intersect for the best interests of both parties. A necessary step to enabling this analysis is the willingness to examine potential oppression placed on transpeople by anti-inclusion feminists, and vice versa.

GenderPAC, a Washington, D.C.-based gender rights group, provides a model for ending gender-based oppression. Their mission statement reads:

> The Gender Public Advocacy Coalition (GenderPAC) works to end discrimination and violence caused by gender stereotypes by changing public attitudes, educating elected officials and expanding human rights. GenderPAC also promotes understanding of the connection between discrimination based on gender stereotypes and sex, sexual orientation, age, race, class. (www.gpac.org)

GenderPAC combines various perspectives of feminist, class, racial, age and queer consciousnesses to work towards ending gender-based oppression. GenderPAC works hard to show that gender-based oppression affects all members of the human community. They accomplish this without appropriating the feminist movement or any other oppressed group, instead creating their own model for change.

However, GenderPAC's efforts are not without criticism. While feminist groups such as the National Organization for Women (NOW) have embraced and collaborated with GenderPAC, trans-activist groups have raised concerns about GenderPAC's practices and trans inclusion. According to GLBTQ Social Sciences, "In early 2001, several transgender activists drafted an open "letter of concern" to GenderPAC, expressing their consternation over the organization's perceived mainstreaming and disconnection from the trans community."[13] This issue became particularly salient during the writing of this article when GenderPAC's Executive Director Riki Wilchins (who reviewed an earlier version of this article) requested that the description of GenderPAC as a "trans-activist group" be changed to "gender rights group." Wilchins stated that GenderPAC "does not want to be known as a transgender group and we are trying to get away from that association."[14] Wilchins has responded to criticism about this intentional distance by arguing that GenderPAC serves all people who transgress gender norms, including transpeople, and that gender rights benefit everyone.

While the intentions are obviously positive and likely politically motivated, this presents another situation where transpeople are placed at a lower priority to benefit a larger group of people. In this case, GenderPAC uses transpeople to benefit a wider community without necessarily returning that benefit to the trans community. One might also question the correlation between GenderPAC's distance from trans-activism and embracement by feminist groups. GenderPAC is obviously not without its faults, such as this distance from the trans community. It does however offer a foundational working model of how feminism and trans-activism can coexist towards a common goal of working towards ending gender-based oppression. This model could be particularly useful in the creation of future organizations and movements aimed at ending gender-based oppression.

CONCLUSIONS

Transwomen and other gender-variant people often find their legitimacy in feminist spaces challenged, in no small part because they are perceived as easy targets. By questioning someone's overall legitimacy in a space, they are proclaiming that this identity is alien to the movement and therefore perspectives represented by this person are extemporaneous. This is particularly effective in space that is defined entirely by gender politics, such as the feminist movement. There is not a simple solution to addressing trans-inclusion in the feminist movement, particularly given the threat of a potentially necessary paradigm shift in feminist theory and ideology. As it stands, feminism as a movement does stand to lose ground by including trans-identities and related issues as a part of its politics. It is much more likely that a movement based entirely on the oppression of 51% of the world's population stands a better chance at success than one that includes the interests of a significantly smaller, more marginalized, and socially unacceptable group of gender-variant people.

However, as the GenderPAC model demonstrates, there are inclusionary tactics available. Even with such working models available, there is no easy solution for feminism. Ultimately it comes down to the question of what is more important to the feminist movement as a whole–seeking the end of oppression of women at the cost of becoming oppressors to others, or taking a substantial risk to reevaluate theory, decentralize woman as the only oppressed gender, and as such ensure that feminism does not become a part of the negative force it is working against.

Sadly, the current politics seem to dictate that the protection of the feminist movement is more important than the oppression of transpeople by anti-inclusion feminism. As the best possible example of feminist consciousness, feminists need to actively work to deconstruct feminist oppression of transpeople. Both feminists *and* transpeople need to continue the dialogue of trans-inclusion and what this means for the feminist movement and trans-activism. This type of work has been done by the feminist movement before, and it is all the stronger because of it. Surely, with the right motivations and dedication, only the same can happen again.

READING QUESTIONS

1. This article suggests that feminists who are against trans-inclusion are in large part motivated by fear. In particular, fears that if trans identities are proven to have biological roots, that this could lead to the dismantling of feminist theory and could potentially discredit the feminist movement. What preemptive steps could feminists take to ensure that such a discovery would not affect the feminist movement?

2. QuestioningTransgender.org states that gender expression will be eliminated when there is no longer sex-based oppression, and that celebrating gender expression means celebrating oppression. Are socially constructed sex roles and gender identities separate entities, and how do we know? Can socially constructed sex roles and gender identities exist separately?

3. What would be some of the pros and cons to feminism broadening its range to include trans identities as a part of its activism? What would be some of the hurdles in implementing such a paradigm shift? What might be some specific steps transpeople could take to assist in this shift? What might be some specific steps anti-inclusion feminists could take to assist in this shift?

4. How does an identity of oppression inhibit feminism's goals? What other communities might be affected by this practice? What would be some specific ways for feminists to move away from this oppressive philosophy?

5. *Essay assignment:* Michigan's Womyn's Music Festival is one of the most prominent and highly contested examples of anti-inclusion feminism. Review the contents of *www.camptrans.org, www. michfest.org, http://questioningtransgender.org/support.htm*, and *http://eminism.org/michigan/faq-intro.html*. What are the arguments brought forth by each side on why transwomen should or should not be allowed on the land? If you were a mediator be-

tween these two parties, with the goal of creating a livable compromise for both sides that valued each of their philosophies, what solutions would you suggest? How do you think that your solutions would be received?

NOTES

1. Notes on terminology: The term 'trans' in this text denotes a person whose gender identity is not congruent with their biological sex. This term is preferred to 'transgender' or 'transsexual' because it does not inherently assume surgical or hormonal status/desire. It is also a purposeful move to include people whose identities do not necessarily fit within the constraints of the more' commonly used terms 'transgender' or 'transsexual' (or who do not feel comfortable with these particular labels). In this piece, the term 'gender diverse' also refers to people whose gender identity is not congruent with their biological sex. This term is purposely chosen, because it is not (currently) affiliated with the terminological border wars within the trans community, and does not necessitate a personal claiming of a trans-specific identity. 'Transpeople' is used to refer to all people who *claim* a trans identity, regardless of the direction of their transition. Further, the term 'transwoman' is used to respectfully refer to a person who was born biologically male and lives as a woman/feminine person, and the term 'transman' is used to respectfully refer to a person who was born biologically female and lives as a man/masculine person. The term 'cisgendered' is used [instead of the more popular 'gender normative'] to refer to people who do not identify with a gender diverse experience, without enforcing existence of a "normative" gender expression. Also of importance to note is that the text views 'gender' and 'sex' as two distinct and separate terms. Herein, gender refers to a person's felt sense of identity and expression, and sex refers to the biological assignment of male, female (and sometimes intersex) at birth based on anatomy (and/or chromosomal arrangements). For more information on these terms and other trans terminology, see: <www.trans-academics. org/LGBTQITerminology.pdf>.

2. The terms 'feminist space' and 'women's space' are used as separate and distinct terms in this text. Feminist space is used to denote a space (physical or metaphorical) that *purposely* exists based on a presumption of feminist consciousness or activism. Women's space refers to a space (physical or metaphorical) that does not center on feminist consciousness, rather *incidentally* centering on the existence of woman dominated or exclusive space.

3. Quoted from the homepage of QuestioningTransgender.org's Website.

4. For example: "I went to women's lib meetings for a while, one MTF stated in 1971, and was getting really into it until some woman wearing an army uniform walked up to me and said that I should take off my false eyelashes and not expose my breasts so much" (Meyerowitz, 259).

5. This was also true of femme lesbians, who were also viewed as traitors for embracing the feminine.

6. See referenced works by Butler, Califia-Rice, and Cromwell.

7. The specifics of the HBSOC are available online at: <http://www.hbigda.org/ soc.cfm>.

8. For definitions of these terms, see Green and Peterson, 2004. LGBTQI Terminology Sheet. Available online at: <www.trans-academics.org/LGBTQITerminology.pdf>.

9. See the Intersex Society of North America (ISNA.org) for a thorough analysis and intersex perspective on sexual assignment surgeries for intersex "conditions." For a more in depth analysis see Dreger AD (Ed.) (1999). *Intersex in the age of ethics.* Frederick, MD: University Press Group; and Kessler SA (1998). *Lessons from the Intersexed.* New Brunswick, NJ: Rutgers University Press.

10. A comprehensive listing is beyond the scope of this paper. However, for a sampling refer to Califia, Patrick (2003). *Sex Changes: The Politics of Transgenderism,* 2nd Ed. San Francisco, CA: Cleis Press; Cromwell, Jason (1999). *Transmen and FTMs: Identities, Bodies, Genders, and Sexualities.* Champaign, IL: University of Illinois Press; Diamond, Morty (2004). *From the Inside Out : Radical Gender Transformation, FTM and Beyond.* San Francisco, CA: Manic D Press; Green, Jamison (2004). *Becoming a Visible Man.* Nashville, TN: Vanderbilt University Press; Prosser, Jay (1998). *Second Skins.* New York: Columbia University Press; Queen, Carol and Schimmel, Lawrence, eds. (1997). *PoMoSexuals: Challenging Assumptions about Gender and Sexuality.* San Francisco, CA: Cleis Press.

11. Note that the internal contradictions of the first two statements are those of Janice Raymond, and not this author. As there are trans-positive authors such as Calfia and others who engage directly with the text, textual analysis is not included here.

12. See Gender.org/remember for a complete listing of and details about the exceptionally violent deaths of transpeople who were targeted for gender diverse related hate crimes.

13. See: <http://www.glbtq.com/social-sciences/gender_public_advocacy.html>.

14. Personal verbal correspondence with author in December 2005.

I Don't Know Who I Am:
Severely Mentally Ill Latina WSW
Navigating Differentness

Sana Loue
Nancy Méndez

SUMMARY. We examine interviews from a qualitative study designed to examine HIV perceptions, risk, and risk management among Puerto Rican women who have sex with women (WSW) and who also have been diagnosed with major depression, bipolar disorder, or schizophrenia. These women's stories challenge both the lesbian and the Latino communities to reexamine how and why they claim individuals as their own and they similarly challenge professional communities, including

Sana Loue, JD, PhD, MPH, is Professor in the Department of Epidemiology and Biostatistics of Case Western Reserve University School of Medicine, Cleveland, Ohio, and is the director of the university's Center for Minority Public Health. Dr. Loue holds advanced degrees in medical anthropology, epidemiology, and law. Dr. Loue's areas of research include the development of HIV prevention interventions for various marginalized communities.

Nancy Méndez is a Research Assistant in the Department of Epidemiology and Biostatistics in the School of Medicine at Case Western Reserve University.

Address correspondence to: Sana Loue, Department of Epidemiology and Biostatistics, School of Medicine, Case Western Reserve University, 10900 Euclid Avenue, Cleveland, OH 44106-4945 (E-mail: Sana.Loue@cwru.edu).

[Haworth co-indexing entry note]: "I Don't Know Who I Am: Severely Mentally Ill Latina WSW Navigating Differentness." Loue, Sana, and Nancy Méndez. Co-published simultaneously in *Journal of Lesbian Studies* (Harrington Park Press, an imprint of The Haworth Press, Inc.) Vol. 10, No. 1/2, 2006, pp. 249-266; and: *Challenging Lesbian Norms: Intersex, Transgender, Intersectional, and Queer Perspectives* (ed: Angela Pattatucci Aragón) Harrington Park Press, an imprint of The Haworth Press, Inc., 2006, pp. 249-266. Single or multiple copies of this article are available for a fee from The Haworth Document Delivery Service [1-800-HAWORTH, 9:00 a.m. - 5:00 p.m. (EST). E-mail address: docdelivery@haworthpress.com].

HIV educators, health researchers, and medical care providers, to develop effective HIV prevention programs and counseling approaches that facilitate patient/client self-disclosure and consider cultural and contextual barriers to both self-disclosure and the provision of services. *[Article copies available for a fee from The Haworth Document Delivery Service: 1-800-HAWORTH. E-mail address: <docdelivery@haworthpress.com> Website: <http://www.HaworthPress.com> © 2006 by The Haworth Press, Inc. All rights reserved.]*

KEYWORDS. Sexual identity, lesbian mental health, lesbian marginalization, Latina lesbians, HIV/AIDS, severe mental illness

This is a snapshot of the lives of eight women, marginalized from both larger, mainstream society and from the smaller communities and groups in which we live, perhaps labeled as *different*, so different, in fact, as to be considered deviant (Becker, 1973; Lemert, 1951). These are women who struggle with minority status as Puerto Ricans; with female status in a community dominated by men and subtle and not-so-subtle demands for women's conformity to a subservient role in male-female relationships; with intimate relations with other women in a heterosexist society; and with severe mental illness (bipolar disorder, major depression, or schizophrenia) in a world that demands full functioning and competence in all phases of life, at every moment. These women can be thought of as queer[1] from multiple perspectives; they struggle each day to appear normal, despite the many dimensions of their *differentness*.

The women's stories represent a sample of the stories that come from a qualitative study based in six counties of northeastern Ohio and San Diego County in California that is examining HIV perceptions, risk, and risk management among Mexican and Puerto Rican women who have diagnoses of major depression, bipolar disorder, or schizophrenia. The eight women presented here reside in northeastern Ohio. In addition to participating in interviews, each woman gave permission to be shadowed, that is, accompanied by an interviewer for a period of 100 hours over two years. The women who are portrayed here have each been shadowed between 4 and 56 hours (mean 18.5 hours), from the time of their enrollment in the study to date.

Several themes are evident across the women's experience. First, many of them have experienced abuse in their relationships with men, as children and later as adults. Their experiences of abuse are often interwoven with themes of substance use, multiple sexual partners, and the increased risk of HIV that accompanies unprotected sex and substance use. For some of these women, the abuse that they have suffered in their relations with men led them to conclude that intimate relationships are best pursued with women. These relationships with women have led many of them to question their own identity in their attempts to be comfortable with themselves and to remain accepted within their social and family circles. All of the women have attempted and continue to attempt to find an explanation for why they suffer from mental illness and to identify a way in which they can manage their disease. Many have ultimately attributed control of their illness to God. Finally, all of these women struggle with their differentness.

ABUSE, SEX, AND HIV RISK

The level of violence that these women have experienced in their lives is sobering. Six of the eight women suffered severe physical or sexual abuse as children, while one-half of them currently are experiencing partner violence.

María often cuts herself "to make sure that [she is] still alive." She was raped by her father and physically abused by her mother when she was younger and continues to be abused by her brothers. She related,

> I was raped when I was five years old by Rafael, Anthony, and my father. I never said nothing until I was seven. It's like a Lifetime movie. It got to the point I started hallucinating of him, my father.

María lives with her current boyfriend, who is also abusive.

> One minute he is the best boyfriend and the other minute his hands are on me. Besides him hitting me he is all right. He twisted my breast, it hurt so bad I cried. He punched me so I smacked him. I said, what the fuck?

Susanna was also sexually abused by her father and often has flashbacks of the incidents. She considers herself a very sexual person and is frustrated by her husband's impotence. Her first sexual experience with a girl occurred when she was 11 years old. At that young age, she was also having sex with four married men. Her feelings about her sexual experiences are interwoven with her religion-oriented hallucinations.

> I've slept with a lot of people and don't want to cheat on my husband. I never used anything [condom] with the men I slept with. The other night I had a dream I was having sex with a plastic demon. In the name of God I am going to get the sexual demons out of my house. I can smell them and see them when they are inside my house. People sometimes bring sexual spirits [demons] with them in my home; I can see them in people. I am very careful who I bring in my house. I eventually gets them all out of the house con ayuno y oracion [prayer and fasting]. God always takes care of me at all times and protected me from getting AIDS. Back then I slept with men because I didn't know sex was a bad thing. We [she and husband] use each other's hand for pleasure. The women I slept with were friends. I wonder how it would be to be with a woman now. Women understand better than men. . . . A lot of black angels are walking around. The black angels are lost souls in the streets. They walk the streets looking for forgiveness because they want to live. Reincarnation, that's the reason why we have to also pray for the dead. A lot of people go to see psychics and God doesn't like that. The souls are people who are dead, who enter animals' bodies like a body of a cat to walk the streets. Everyone wants to be saved and sometimes they are stuck in the streets so the demons possess them. If we pray for the dead God will hear us. Kids who die and are under nine years old go straight to heaven because they haven't sinned yet. The dark angels also use people's bodies, they enter people's bodies who have mental illnesses, people who are in psychiatric facilities, and homeless people. They also enter bodies emotionally abused and sexual offenders. Las almas estan torturadas [The souls are being tortured]. I don't judge people but prostitutes have a evil spirit of sexuality, it's a bad spirit. I struggle with this spirit.

Susanna talks frequently about the orgies that she attended. Like several of the other women, she is aware of her HIV risk, but refuses to be tested. Susanna is especially worried that she may have had sex with an HIV-positive person and believes it is likely that she will become infected if she is not already. Susanna's brother, who became HIV-infected due to injection drug use, died of AIDS.

Marta was raped by her father when she was 18. Since that time, she has heard his voice speaking to her. She was also beaten by both parents. She has used crack, cocaine, marijuana, and alcohol in the past, and continues to use alcohol and marijuana. She has frequent hallucinations and has made many suicide attempts that have necessitated hospitalization.

Marta has never been married but has had numerous male and female sexual partners. She has been arrested on at least one occasion for harassment and has difficulty controlling her own violent behavior. She is very worried that she will become infected with HIV if she is not already, and believes that she probably has had sexual relations with someone who is HIV-positive. She refuses to be tested, however, because she does not believe that the results are really kept confidential or anonymous. Marta does not want to use condoms during sex because she believes that it means she cannot trust her partner. She never asks a partner, male or female, if they have been tested for HIV. She does not utilize the local legal needle exchange program because she finds it embarrassing to ask for a clean needle.

The women's reports of incest as children and subsequent sexual relationships with multiple sexual partners are similar to patterns reported in the literature. Studies of women with histories of incest have often reported sexual promiscuity as one of the sequelae (e.g., Lukianoiwicz, 1972; Maisch, 1972; Medlicott, 1967; Riszt, 1979). The validity of these findings is uncertain in view of significant methodological problems associated with the studies. Many suffer from selection bias; they include only women drawn from clinical and forensic samples. Societal standards were significantly harsher for female sexual behavior at the time the studies were conducted than they might be today. Additionally, use of the term *promiscuous* to describe resulting behavior serves only to cast moral judgment, rather than to describe in a meaningful way the specific outcome. Consequently, it is unclear what type or frequency of behavior is actually encompassed within that term.

Although disturbing, the level of violence that we see in the lives of these women is not surprising. Existing research indicates that severely mentally ill women may be at increased risk for partner violence, perhaps resulting from impaired judgment, poor reality testing, and difficulties associated with planning that comprise many mental illness diagnoses (Goodman et al., 1997). Partner violence committed against women suffering from severe mental illness has been found to further exacerbate their behaviors or symptoms (Campbell, 2002; Weingourt, 1990).

Research findings also suggest that Latinas may be at elevated risk for partner violence. Hispanic women have been found to be at greater risk of physical violence during marriage compared to women of other ethnic groups (Straus and Smith, 1990). In comparison with non-Hispanic white women, Hispanic women are more likely to have been the victims of violence for a longer period of time (Gondolf, Fisher, and McFerron, 1988; Torres, 1991). Several research groups have reported that factors such as immigration status, prejudice, a lack of English proficiency, and the lack of emotional support resulting from separation from extended families may contribute to the abuse (Ho, 1990; Perilla, Bakeman, and Norris, 1994). With respect to Puerto Ricans, it has been found that Puerto Rican husbands are ten times more likely than Cuban husbands to assault their wives (Kantor, Jasinski, and Aldarondo, 1994). Compared to non-Hispanic whites, Cubans, and Mexican-Americans, Puerto Ricans have the highest rate of cultural approval of wife assaults (Kantor, Jasinski, and Aldarondo, 1994). Given this, it is not surprising that Puerto Ricans have been found to be more tolerant of wife assault compared to several other groups, including non-Hispanic whites (Kantor, Jasinski, and Aldarondo, 1994). The reasons for this culturally specific elevated risk remain unclear.

FINDING ONE'S IDENTITY

Study participants have struggled to find their own identity. Some reported that their decision to have relationships with women stemmed from multiple episodes of abuse inflicted by male relatives and male partners.

Blanca, who reported a history of abuse and rape by her father as a small child, talked about the "last straw" that turned her from men.

> When he cheated on me, I broke everything I had. During that time we were arguing a lot because I knew that he was cheating. He wasn't coming home until 3 or 4 in the morning. He admitted he was fucking with someone. We were fighting in the kitchen and I threw a kitchen gadget at him and cut him. Right there he knew that I really loved him. I know the girl that he was cheating with. Her name is Gilda. I called her and she admitted sleeping with him. She said to me, "If it makes you feel better, we use a condom." After I confirmed it, I threw all his shit outside and got real sick and ended up in the hospital.

As a result, Blanca, says, she is "into women" and will no longer be with men: "I don't know if you know but I am interested in girls now."

Although Christina has indicated that she wants to stay with women from now on and has had multiple relationships with women, she does not identify as a lesbian.

> I think that I am gay, not a lesbian. Well, a lesbian is women who are butch. A gay person is a person who is attracted to females that dress like men. If I saw a cute guy I would say something. But I do not see myself sexually with a man.

Christina describes her relationships with women as a matter of choice, rather than a biological imperative.

> I dated my first girlfriend in 2000. I was confused back then so I eventually went back to men. But now I have made up my choice to stay with women. I don't plan to go back to men. I prefer sticking with women. Because females know what you want emotionally, physically and sexually. [My girlfriend] always rubs my hair for me to go to sleep. . . . I really enjoy being with [my girlfriend]. When we go out we have so much fun. We hold hands everywhere we go.

A number of studies have examined sexual orientation among women with incest histories. Many report that a proportion of the women who experienced incest have had same-sex sexual experiences and/or self-identified as lesbian (Finkelhor, 1980; Herman, 1981; Meiselman, 1978). Other studies of the histories of lesbians have found that a minority had suffered incest (Simari and Baskin, 1982). However, no causal link has been established between incest, whether by a same-sex or opposite-sex perpetrator, and later sexual orientation or partner preference.

MANAGING MENTAL ILLNESS

Many of the women have turned to God in an attempt to manage their symptoms. Unfortunately, none of these women have received support–emotional, financial, or physical–from their churches. The solace that they find in these explanations and in their faiths exists on a purely personal level in their relationship with God and His agents, as the women each define it for themselves.

Dara, like her parents, was born in Puerto Rico. She has lived on the mainland for approximately 30 years. Dara suffered severe physical abuse as a child and, as a consequence of her illness, she is unable to maintain employment and subsists on social security payments. She has said of her illness, "I don't control it; it controls me." She refuses to take medication that has been recommended for her illness. Dara believes that God does not give anyone anything that they cannot handle, and that God knows ahead of time everything that will happen to a person. In fact, Dara has "asked God to be [her] meds."

Blanca was also raped and physically abused as a child by her father. She attributes her illness to having "lots of responsibility" with her children. She believes that God is more likely than medicines to cure her illness.

Like Blanca, Lydia also believes that her illness is more likely to be cured by God than by medicine. She has heard voices since she was quite young and has been hospitalized several times. She has explained how she believes she developed schizophrenia.

> I know that my condition was not something I inherited. I developed this condition because of a burn I suffered when I was three years old. Well, I was three years old and I was really sick with a fever. My mom bathe me with alcohol and alcolado. My hair was soaked in it. My brother Tito was playing with a toy gun that sparked. He did it next to me and my hair lit up in fire and it burned the whole side of my face. I was 6 years without being able to grow hair. That is why my brother Tito and I are so close cause he always felt terrible about that incident.

Susanna experiences frequent hallucinations of angels, demons, and disfigured people. She once believed that she could fight the devil and win, but now believes that God holds ultimate power. The angels protect and comfort her.

> I used to sometimes think I had power over him [the devil] and would want to fight him. I always thought I could fight him and win. The angels used to talk nice to me and help me a lot. . . . I liked the angels very much, they tell me beautiful things. I still sees them. Ellos viven aqui en mi cuarto. (They live in my room.) They have names and they live with me and protect me from evil. Four angels live with me. They are named Diamante, Querubil, and Safiro. If the angels leave I'll feel lost. They are always in my

room watching me sleep. They are beautiful, big wings, and big. I want to be just like God because she has power. I wanted to help people with their problems. When I speak to the angels in the room my daughter always asks who I am talking to. People's problems are frustrating and I cannot help people the way I thought I could because I feel inferior to God. I have to learn that God is in charge of all the powers and not me. Yo era dominante, papa dios tiene el poder yo no (I was dominant, father god has all the powers not me).

Several years ago, Susanna believed that she was the Virgin Mary and had to be hospitalized. She lives on her social security checks and is very conscientious about taking her medications. She has made several suicide attempts, once by drinking rubbing alcohol and another time by swallowing mercury.

STRUGGLING WITH DIFFERENTNESS

What is clear is that all of these women live lives characterized by enormous complexity and multiple dimensions of differentness. Each experiences this sense of differentness and accompanying feelings of ostracism and marginalization. For some, it is rooted primarily in their sexual identity. We see, to varying degrees, attempts by the women to blend into what they perceive as *normal* life.

María is falling in love with another woman, but is tormented by such feelings: "I'm struggling not to be gay, I want to be straight. I have issues with God." The confluence of identities has resulted in significant confusion for Maria. She lamented,

> I'm so confused. I don't know who I am anymore. I look at myself in the mirror and ask myself who are you? When I look at myself in the mirror I asked why?, What?, How? I don't know who I am and what I want. Robert [boyfriend] gets mad at me because he asked me if I ever thought of women and I said yes. I told him every woman can look at another woman and is able to tell if she is pretty or not. Then he asked me if I ever thought of women in a sexual way and I said yes. He then got real mad. I love Robert but I'm confused about my sexuality.

María wants "the American Dream," but wants to marry a woman as well:

> I wanted to marry a girl. This girl a while back. She was pregnant at the time. I stayed with her the entire pregnancy. I used to rub her belly. Then we broke up . . . I'm confused about my sexuality and I want to be different than my family. I have always tried to be opposite of my family. I want to make it to the top because that's the American dream. To be a family . . . to have a mother and father. I was ashamed to say my mom was gay. I needed a father a role model. My mom didn't teach us. I want my child to have a father.

María is also aware of the negative reaction that she would suffer from the community if her sexual orientation were to become known. She related an incident that occurred with her girlfriend:

> I used to go to a non-denomination church but the pastor told her [girlfriend] she should wear a dress because she was a woman not a man. She said she was not allowed in dressed like a man.

In despair, María told her boyfriend, "I feel like killing myself, I am thinking how much I hate me."

Ana regularly hears voices that tell her to kill herself. She has thrown herself in front of moving cars and almost jumped off of a bridge. She started drinking heavily at the age of 13 and says that drinking helps her to think of ways to kill herself. Her only romantic relationship with a man was with her husband; her marriage ended due to his infidelity. Ana now considers herself *butch* and is in love with two women who, although related to each other, are unaware that Ana loves them both. Ana says that she feels like a man and prefers younger women because the sex is better. One of her girlfriends, who is also Latina, does not want anyone to know about their relationship because she does not want to be thought of as a lesbian. Ana feels that no one, including her mother, knows that she likes women:

> Mom doesn't really know. I think mom knows but she doesn't recognize it. Because of the way I look I'm a soft butch. I'm not a femme. I don't wear skirts and stuff like that. I have a cousin that is a lesbian and they treated her bad. They abandoned her. That's why Marta doesn't want anyone to know she likes me.

Lydia "married" her previous girlfriend while they were living in Texas. Despite her relationships with men, she asserts that she has always been *alegre*, or gay. But she says wistfully,

I can't be gay here. My family doesn't like it. When I lived in Puerto Rico and got married, I still kept my women lovers. The town always gossiped about me. I did eventually tell my mom and brothers and they said I was disgusting. . . . They think it is not normal, it is a sin. I am trying to have a good life, trying to be with a man, but I don't like to have sex with him so I tell him it is the drugs so he will not feel bad.

The reluctance of these women to self-identify as lesbian may be attributable to cultural and community expectations of women, concomitant constraints on their behavior, whether heterosexual or lesbian, and the women's fears of the potential consequences should they violate these mandated behavioral scripts. Sex with men is highly stigmatizing behavior within lesbian communities (Richardson, 2000), and woman-woman sex may be equally marginalized within Latino communities (Ward, 2004; cf. Román, 1995). The culturally constituted concept of *hembrismo*, or femaleness, that simultaneously encourages women to be submissive to men while aggressively pursuing the achievement of their personal goals may underlie the decision of some of these women, such as Lydia, to situate their *queerness* (as they are perceived by others) through the queerness of a male partner, who may know of the woman's sexual attraction to other women and express a desire to participate in or watch their sexual play. Their continued participation in a sexual relationship with a male allows them to maintain a safe and acceptable identity within their Latino community, while they simultaneously pursue a more satisfying emotional and sexual relationship with a woman. The women's fears of marginalization and ostracism are grounded in reality, as evidenced by harsh comments made to María by her pastor, the reaction of Lydia's family members, and the abandonment of Ana's lesbian cousin.

Regardless of the dynamic that underlies these women's motivation to maintain secrecy around their same-sex relationships, this decision comes at a terrible cost. All of these women have located themselves and defined their identities in response to cultural definitions that dominate the lesbian and Latino communities. Christina, for instance, defined herself as gay, but not lesbian, because she is not a butch and is attracted to women who look like men. Despite the variations in the construction of their identities in response to these external demands, all of these women appear to be at increased risk of HIV infection. Women-to-woman transmission of HIV is possible (Diamant, Lever, and Schuster, 2000; Lesbian AIDS Project, 1994; Morrow and

Allsworth, 2000; Rose, 1993; Stevens, 1994), although it has been found to be relatively rare in the absence of other risk behaviors, such as the sharing of injection equipment or unprotected sexual intercourse with an infected male partner (Bevier, Chiasson, Heffernan, and Castro, 1995; Chu, Buehler, Fleming, and Berkelman, 1990; Chu, Conti, Schable, and Díaz, 1994; Chu, Hammett, and Buehler, 1992; Petersen, Doll, White, Chu, and the HIV Blood Donor Study Group, 1992). Some of the women, like Ana, who now has sexual relations with only women, fail to recognize the small but real risk that attends these encounters and do not use risk reduction practices, believing that the sexual activities that they engage in are not risky, despite their objective risk (Lemp, Jones, Kellogg et al., 1995; Lesbian AIDS Project, 1994; Raiteri, Fora, Gioannini et al., 1994; Stevens, 1994) and/or believing that they are at little or no risk of HIV because, as a social group, lesbians are unlikely to be HIV-infected (Richardson, 2000). Other women, such as Christina, recognize that they are at increased risk for HIV because of their partner choice, but feel constrained in their options in view of the potential consequences. Ultimately, we must confront the reality that these women have known for some time: full disclosure of who they are will likely result in their marginalization and ostracism from both the Latino and lesbian communities, while a failure to disclose similarly results in an inability to obtain critical information, guidance, and support.

Although their mental illness cannot be attributed to the precariousness of their *self-situatedness*, one must ask whether their liminal status between communities and their awareness of their own marginalization may contribute to their inability to progress to both improved health status and a healthier lifestyle. Indeed, the

> awareness of stigma that surrounds homosexuality leads the experience to become an extremely negative one: shame and secrecy, silence and self-awareness, a strong sense of differentness–and of peculiarity–pervades the consciousness. (Plummer, 1995: 89)

The internalization of this stigmatization, variously known as internalized homophobia (Meyer and Dean, 1998) and internalized homonegativism (cf. Hudson and Ricketts, 1980; Williamson, 2000), has been suggested as a factor for participation in riskier sexual behaviors (Williamson, 2000), increased substance use and alcohol consumption (Finnegan and Cook, 1984; Glaus, 1988; Meyer and Dean, 1995), self-mutilation, and suicidality (Hammelman, 1993). Seven of the eight women have disclosed histories of suicide attempts, seven of substance abuse (alcohol and illicit drugs), and one of cutting behavior. María, for instance, clearly stated that she wished to kill herself and she hates herself. The data that we have collected to date is insufficient to examine the possible association between internalized homonegativism and its effect on mental health status.

Some of the women also struggle with a sense of differentness that arises from the reactions of those around them to their mental illness. Lydia aptly summarized the situation that is faced by these women:

> People think that because you have a mental illness that you are crazy. There is a lot of discrimination against the mentally ill . . . "Yo estoy enferma pero no loca" (I may be ill but I am not crazy). There are worser things than having a mental illness like prostitutes and drug addicts.

There is the old saying that just because someone is paranoid, does not mean that they are not being followed. Lydia's mental illness does not prevent her from perceiving–or cause her to perceive–the responses of others. Researchers have documented the negative attributions may be "audiences" to individuals diagnosed with or perceived as having mental illness and the efforts that individuals make to avoid others' recognition of their symptoms (Goffman, 1961, Scheff, 1966, 1974; Szasz, 1960, 1961).

DISCUSSION

As Gloria Anzaldúa states, "Identity is not a bunch of little cubbyholes stuffed respectively with intellect, race, sex, class, vocation, gender. Identity flows between, over, aspects of a person. Identity is a river–a process" (1991: 252-253). This observation is apropos here, as we are witness to the many facets of these women's lives and their struggles to reconcile the multiple barriers to their full expression.

Although previous studies have documented an increased risk of HIV infection among severely mentally ill individuals, attributable in part to multiple same-sex and opposite-sex sexual encounters (Brunette et al., 1999; Cournos et al., 1994; Hanson et al., 1992; Kalichman et al., 1994; Kelly et al., 1992), relatively few studies have examined the nature of those encounters (violence, casual or significant nature of relationship, etc.) (Chandra, Carey, Carey, Shalinianant, and Thomas,

2003; Weinhardt, Carey, and Carey, 1998); none have investigated HIV risk among severely mentally ill women in the context of their sexual identity (rather than the sex of the partner) and perceived cultural expectations. Although our findings relate to a small sample and, as such, are not generalizable, they suggest additional cultural expectations and barriers that should be considered in the designing and delivery of HIV prevention interventions for severely mentally ill Puerto Rican women.

The majority of the women whose situations are related here are not *out* to the their families, their friends, or their healthcare providers. To a large extent, their unwillingness to self-disclose results from a fear of ostracism and abandonment by those who appear to care for them and about them. These women are already highly vulnerable due to their mental illness and the level of violence that is present in their daily lives. The loss of this supportive network, however critical and demanding it may be, could leave the women with nothing. Although some might argue that these women have created their own difficult situations as a result of their choices to self-situate in limine between the dominant values of the lesbian and Latino cultures, such a perspective ignores the politics within each of these cultures and fails to appreciate the reality that these women face on a daily basis.

These women's stories challenge both the lesbian and the Latino communities to reexamine how and why they claim individuals as their own and the motivation that prompts communities that have been ostracized and devalued themselves, such as the lesbian and Latino communities, to treat individuals in a similar fashion. They similarly challenge professional communities, including HIV educators, health researchers, and medical care providers, to develop effective HIV prevention programs and counseling approaches that facilitate patient/client self-disclosure and consider cultural and contextual barriers to both self-disclosure and the provision of services. A failure to acknowledge and modify this dynamic essentially mandates that women such as those who are portrayed here remain perpetually closeted and in limine, with all of the ramifications that adhere to that status.

ISSUES AND QUESTIONS FOR FURTHER CONSIDERATION

1. Many of the research participants portrayed here attempt to understand their mental illness and/or relate to others around them and to their environment by reference to one or more aspects of their religion, for example, good and bad angels, God, etc. How

can their connection to God and/or religion be utilized in a positive way to improve the quality of their lives? To reduce HIV risk?

2. Many churches and faiths have rejected and/or ostracized their members who have sexual relations with same-sex partners. Is it possible to involve such churches in a positive and productive way to assist these women if the women were to be candid about who they are and who they love? Why or why not? How?

3. Many of the research participants reported having been sexually abused as children. Is this related to their mental illness and, if so, how and why?

AUTHOR NOTE

WSW is an acronym for *women who have sex with women*, a designation used in public health research that is focused on changing risky sexual behavior and is not dependent upon constructionist identities such as lesbian or bisexual.

This research is part of a larger study that is examining the context of HIV risk among severely mentally ill Latinas, funded by the National Institute of Mental Health (MH63016).

NOTE

1. The word queer is being used here both in its traditional context applied to sexual orientation and also in a broader framework in which it is a metaphor for marginalization because of difference.

REFERENCES

Anzaldúa G (1991). To(o) queer the writer: Loca, escrita y chicana. In: Warland B (Ed.). *InVersions: Writing by Dykes, Queers, and Lesbians*. Vancouver: Press Gang.

Becker HS (1973). *Outsiders: Studies in the Sociology of Deviance*. New York: Free Press.

Bevier PJ, Chiasson MA, Heffernan RT, and Castro KG (1995). Women at a sexually transmitted disease clinic who reported same-sex contact: Their HIV seroprevalence and risk behaviors. *American Journal of Public Health 85*(10), 1366- 1371.

Brunette MF, Rosenberg SD, Goodman LA, Mueser KT, Osher FC, Vidaver K, Auciello P, Wolford GL, and Drake RE (1998). HIV risk factors among people with severe mental illness in urban and rural areas. *Psychiatric Services (Washington, D.C.), 50*(4), 556-558.

Campbell JC (2002). Health consequences of intimate partner violence. *Lancet 359*(9314), 1331-1336.

Chandra PS, Carey MP, Carey KB, Shalinianant A, and Thimas T (2003). Sexual coercion and abuse among women with a severe mental illness in India: An exploratory investigation. *Comprehensive Psychiatry 44*(3), 205-212.

Chu SY, Buehler JW, Fleming PL, and Berkelman RL (1990). Epidemiology of reported cases pf AIDS in lesbians, United States 1980-89. *American Journal of Public Health 80*(11), 1380-1381.

Chu SY, Conti L, Schable BA, and Díaz T (1994). Female-to-female sexual contact and HIV transmission. *Journal of the American Medical Association 272*(6), 433.

Chu SY, Hammett TA, and Buehler JW (1992). Update: Epidemiology of reported cases of AIDS in women who report sex with only other women, United States, 1980-1991. *AIDS 6*(5), 518-519.

Cournos F, Guido JR, Coomaraswamy S, Meyer-Bahlberg H, Sugden R, and Horwath E (1994). Sexual activity and risk of HIV infection among patients with schizophrenia. *American Journal of Psychiatry 151*, 228-232.

Diamant AL, Lever J, and Schuster MA (2000). Lesbians' sexual activities and efforts to reduce risks for sexually transmitted diseases. *Journal of the Gay and Lesbian Medical Association 4*(2), 41-48.

Finkelhor D (1980). Sex among siblings: A survey on prevalence, variety, and effect. *Archives of Sexual Behavior 9*, 171-194.

Finnegan D and Cook D (1984). Special issues affecting the treatment of gay male and lesbian alcoholics. *Alcoholism Treatment Quarterly 1*, 85-98.

Glaus O (1988). Alcoholism, chemical dependency and the lesbian client. *Women & Therapy 8*, 121-144.

Goffman E (1961). *Asylums*. Garden City, New York: Anchor Books.

Gondolf EW, Fisher E and McFerron JR (1988). Racial differences among shelter residents: A comparison of Anglo, black, and Hispanic battered women. *Journal of Family Violence 3*, 39-51.

Goodman LA, Johnson MM, Dutton MA, and Harris M (1997). Prevalence and impact of sexual and physical abuse. In: Harris M and Landis CL (Eds.). *Sexual Abuse in the Lives of Women Diagnosed with Serious Mental Illness*. The Netherlands: Harwood Academic Publishers, pgs. 277-299.

Hammelman T (1993). Gay and lesbian youth: Contributing factors to serious attempts or considerations of suicide. *Journal of Gay & Lesbian Psychotherapy 2*: 77-89.

Hanson M, Kramer TH, Gross W, Quintana J, Ping-Wu L, and Ashe R (1992). AIDS awareness and risk behaviors among dually disordered adults. *AIDS Education and Prevention 4*, 41-51.

Herman J (1981). *Father-Daughter Incest*. Cambridge, Massachusetts: Harvard University Press.

Ho CK (1990). An analysis of domestic violence in Asian-American communities: A multicultural approach to counseling. In: Brown LS and Root MPP (Eds.). *Diversity and Complexity in Feminist Therapy*. New York: Haworth Press, pgs. 129-150.

Hudson W and Ricketts W (1980). A strategy for the measure of homophobia. *Journal of Homosexuality 5*, 357-372.

Kalichman SC, Kelly JA, Johnson JR, and Bulto M (1994). Factors associated with risk for HIV infection among chronically mentally ill adults. *American Journal of Psychiatry 151*, 221-227.

Kantor GK, Jasinski JL, and Aldarondo E (1994). Sociocultural status and incidence of marital violence in Hispanic families. *Violence and Victims 9*: 207-222.

Kelly JA, Murphy DA, Bahr GR, Brasfield TL, Davis DR, Hauth AC, Morgan, MG, Stevenson LY, and Eilers MK (1992). AIDS/HIV risk behavior among the chronic mentally ill. *American Journal of Psychiatry 149*, 886-889.

Lemert EM (1951). *Social Pathology.* New York: McGraw-Hill.

Lemp GF, Jones M, Kellogg TA, Nieri GN, Anderson L, Withum D, and Katz M (1995). HIV seroprevalence and risk behaviors among lesbians and bisexual women in San Francisco and Berkeley, California. *American Journal of Public Health 85*(11): 1549-1552.

Lesbian AIDS Project. (1994). *Results of the Lesbian AIDS Project's Women's Sex Survey: Final Report.* New York: Lesbian AIDS Project/Gay Men's Health Crisis.

Lukjanowicz N (1972). Incest. *British Journal of Psychiatry 120*, 301-313.

Maisch H (1972). *Incest.* (C. Bearne, Trans.). New York: Stein & Day.

Medlicott R (1967). Parent-child incest. *Australian and New Zealand Journal of Psychiatry 1*, 180-187.

Meiselman K (1978). Personality characteristics of incest history psychotherapy patients: A research note. *Archives of Sexual Behavior 9*: 195-197.

Meyer I and Dean C (1998). Internalized homophobia, intimacy, and sexual behaviour among gay and bisexual men. In: Herek G (Ed.) *Stigma and Sexual Orientation.* Thousand Oaks, California: Sage.

Morrow KM and Allsworth JE (2000). Sexual risk in lesbians and bisexual women. *Journal of the Gay and Lesbian Medical Association 4*(4): 159-165.

Perilla JL, Bakeman R and Norris FH (1994). Culture and domestic violence: The ecology of abused Latinas. *Violence and Victims 9*, 325-339.

Petersen LR, Doll L, White C, Chu S, and HIV Blood Donor Study Group. (1992). No evidence for female-to-female HIV transmission among 96,000 female blood donors. *Journal of Acquired Immune Deficiency Syndromes 5*, 853-855.

Plummer K (1995). *Telling Sexual Stories: Power, Change, and Social Worlds.* London: Routledge.

Raiteri K, Fora R, Gioannini P, Russo R, Lucchini A, Terzi MG, Giacobbi D and Sinicco A (1994). Seroprevalence, risk factors, and attitudes to HIV-1 in a representative sample of lesbians in Turin. *Genitourinary Medicine 70*, 200-205.

Richardson D (2000). The social construction of community: HIV risk perception and prevention among lesbians and bisexual women. *Culture, Health & Sexuality 1*, 33-49.

Riszt K (1979). Incest: Theoretical and clinical views. *American Journal of Orthopsychiatry 49*(4), 680-691.

Román D (1995). Teatro Viva!: Latino performance and the politics of AIDS in Los Angeles. In: Bergmann EL and Smith PJ (Eds.). *Entiendes? Queer Readings, Hispanic Writings.* Durham, North Carolina: Duke University Press, pgs. 346-369.

Rose P (1993). Out in the open? *Nursing Times 89*: 50-52.

Scheff TJ (1974). The labeling theory of mental illness. *American Sociological Review 39*, 444-452.

Simari C and Baskin D (1982). Incestuous experiences within homosexual populations: A preliminary study. *Archives of Sexual Behavior 11*, 329-344.

Stevens PE (1994). Lesbians and HIV: Clinical, research, and policy issues. *Journal of Orthopsychiatry 63*, 289-295.

Straus MA and Smith C (1990). Family patterns and child abuse. In: Straus MA and Gelles RJ (Eds.). *Physical Violence in American Families: Risk Factor Adaptations to Violence in 8,145 Families*. New Brunswick, New Jersey: Transaction Publishers, pgs. 245-261.

Szasz TS (1961). *The Myth of Mental Illness*. New York: Hoeber-Harper.

Szasz TJ (1960). The myth of mental illness. *American Psychologist 15*, 113-118.

Torres S (1991). A comparison of wife abuse between two cultures: Perceptions, attitudes, nature, and extent. *Issues in Mental Health Nursing: Psychiatric Nursing for the 90s: New Concepts, New Therapies, 12*, 113-131.

Ward J (2004). Not all differences are created equal: Multiple jeopardy in a gendered organization. *Gender & Society 18*(1), 82-102.

Weingourt R (1990). Wife rape in a sample of psychiatric patients. *IMAGE: Journal of Nursing Scholarship 22*(3), 144-147.

Weinhardt LS, Carey MP and Carey KB (1998). HIV-risk behavior and the public health context of HIV/AIDS among women living with a severe and persistent mental illness. *Journal of Nervous and Mental Disease 186*, 276-282.

Williamson IR (2000). Internalized homophobia and health issues affecting lesbians and gay men. *Health Education Research: Theory and Practice 5*(1), 97-107.

Index

Abuse, Latina lesbians and, 251-252
Allen, Paula Gunn, 83
Althusser, Louis, 95
Androcentrism, 183-184
Antiontologizing sex, 31
Antiontology, 27
Australian WOMAN Network,
202-212

Barthes, Roland, 80
Beauvoir, Simone de, 75
Beemyn, Brett, 160
Bem, Sandra, 183
Better Than Chocolate, 216-228
Bisexuality, 78-79
Body
    one-sex model of, 184-185
    Spherical Characterization Model
        of, 186-190
    transgender individuals and,
        185-186
Bolonik, Kera, 92
Bornstein, Kate, 29,31,32,159,160,190
Bottoming, 64-65
Boyd, Nan, 189
Boys Don't Cry, 94,95,97,98-99
Brandon, Teena Renae. *See* Teena,
    Brandon
Brant, Beth, 160,161-162
Brown, Wendy, 95-96
Bruss, Elizabeth, 94-95
Butch codes, 77-78
Butches, 22,90,91-92
Butler, Judith, 26,30,32-33,
    77,78,81,88,95,237

Califia, Patrick, 232
Cameron, Loren, 34,47
Chase, Cheryl, 6
Codes, butch, 77-78
Coherentist assumptions, 91
Collis, Rose, 113-114
Cook, Blanche Wiesen, 94
Coughlin, Colleen, 12
Counteridentification, 89-93
Cromwell, Jason, 46
Cronson, Charlotte, 232
Cross-dressing, 29-30,77
Curriculum, LGBTQ, 155-157

Dale, Elizabeth, 12
Davies, Carol Boyce, 79
De Lauretis, Teresa, 100
Detloff, Madelyn, 22
Differential consciousness, 81-84
Dolan, Jill, 112
Dress, 30

Eliason, Mickey, 160
Erikson, Erik, 65
Essentialism, 20
Essentialist lesbian feminism, 183
Eugenides, Jeffrey, 193-196

Feinberg, Leslie, 64,89,98,160,161
Female-to-male (FTM) trans activists,
    1-12
Female-to-male (FTM) transsexuals,
    18,19-20,90,91-92

Feminism
  identity of oppression and, 242-243
  trans identities and, 232
Feminist theory, 233-234
Femme identity, 77-78
Femme lesbians, 14n13,74-84
Foucault, Michel, 93

Garber, Marjorie, 29-30,31,32
Gaydar, 74-75,84
Gay vibes, 74-75,79-81,84
Gender, 29-30
  in Native societies, 126
  policing, 94-101
Gender acquisition, 32
GenderPAC, 244-245
Gender performativity, theory of,
    26,32-33,77
Gender transitioning, 25-26
Green, Jamison, 93

Halberstam, Judith, 18,22,25,26,30,
    88,94,160,161
Hale, C. Jacob, 18,26,28,92,94
Halley, Janet, 91
Harry Benjamin Standards of Care
    (HBSOC), 235,236
Hill, Darryl, 45
HIV prevention, queer-identified youth
    and, 64-65
HIV risk, Latina lesbians and, 252-254
Homonormativity, in lesbian
    communities, 9-10
HONOR Project, 127-128

Identitarian knowledge formations,
    25-32
Identity
  inclusive model of, 186-190
  lesbian, 182-183
  of oppression, feminism and,
    242-243

If These Walls Could Talk Two,
    116-118
Intersectional individuals, 9
Intersex females, 13n7
Intersex individuals, 9,45

Latina lesbians, 250-263. *See also*
    Lesbians
  abuse and, 251-252
  finding one's identity and, 254-255
  HIV risk and, 252-254
  managing mental illness and,
    255-257
  struggling with differentness and,
    257-261
  violence and, 253-254
Lesbian communities
  defined, 19
  homonormativity in, 9
  removing exclusionary practices
    from, 182-183
  same-sex marriage and, 8-9
  trans inclusion and, 8-10
  transsexuals and, 19-22
Lesbian feminism, 90-91
  essentialist, 183
Lesbian identity, 182-183
Lesbianism, television movies and,
    111-114
Lesbian normativity, challenging,
    overview of, 1-12
Lesbians. *See also* Latina lesbians;
    Two-spirit women
  authenticity and, 7-8
  femme, 14n13,74-84
  legitimacy and, 7-8
Lesfest, 202-212
LGBT (Lesbian, Gay, Bisexual, and
    Transgender) health, 44
LGBTQ (Lesbian, Gay Bisexual,
    Transgender, and Queer)
    curriculum, 155-157

Male-to-female (MTF) transsexuals, 5
Marginality, 83
Marriage, restrictions on, 13n11
Martin, Biddy, 77
Masculinity, pain and, 93
Mason-Schrock, Douglas, 189
Michigan Womyn's Music Festival,
    183-184
Middlesex (Eugenides), 193-196
Miller, D. A., 95
Minkowitz, Donna, 28
Munt, Sally, 96
Myth, 80

Namaste, Ki, 160,161
Nanda, Serena, 160,161-162
Native societies, gender in, 126-127
Nestle, Joan, 75
Newton, Esther, 89,90,91
Noble, Jean Bobby, 93
Normal, 121-122

One-sex model, of body, 184-185
Oppression
    gender-based, 243-245
    identity of, feminism and, 242-243
Orlando (Woolf), 190-193
Outerbridge, Peter, 218

Pain, masculinity and, 93
Pleasure, positive articulation of, 93
Policing gender, 94-101
Public fantasies, 100-101

Queer, regendering, 30-31
Queer identification, 45,64
Queer-identified youth
    health care providers and, 65-66
    health services and, 46-47
    HIV prevention approaches for, 64-65

study of, 47-49
Queer individuals, 9
Queer sexuality, 81
QuestioningTransgender.org, 239-241

Rapping, Elayne, 109-110
Raymond, Janice, 25,232,234-239
Rednour, Shar, 77
Reid, case study of, 56-64
Repressive State Apparatuses, 95
Román, David, 160
Rothblum, Esther, 6,12
Rubin, Henry S., 20-21,46-47,90

Samantha, case study of, 49-56
Same-sex marriage
    lesbian communities and, 8-9
    views on, 15n14
Sandoval, Chela, 81-82,83
Scarry, Elaine, 96
Sedgwick, Eve Kosofsky, 26,27
*Serving in Silence: The Margarethe
    Cammermeyer Story,*
    114-116
Sexed ontology, 21,27
Spherical Characterization Model, of
    body, 186-190
Spirituality, lesbians and. *See*
    Two-spirit women
Stone, Sandy, 235
Stone Butch Blues, 94,95,97,98-99
Stryker, Susan, 22
Substantiation, 96-97
Subversion, 77-78

Teaching transgender, 152-175
Teena, Brandon, 28-29,94,97-98
Television, queers on, 108-109
Television movies, 109-122
    lesbianism and, 111-114
Thompson, Petty, 218

Trans, 13n5,30-31
Transgender, 22-25
    identifying as, 22-23
    student outcomes, 169-172
    teaching, 152-175
    testimonial, 164-169
    transformative power of teaching,
        172-175
    as umbrella term, 22
Transgender-identified youth
    study of, 47-49
        Reid, 56-64
        Samantha, 49-56
Transgender individuals, 9,24
    body and, 185-186
    pain and suffering of, 24
Trans identities, feminism and, 232
Trans inclusion, issue of, 8
    lesbian communities and, 8-10
Transitioning, 32
Transmale-identified youth, health
        services and, 46-47
*The Transsexual Empire: The Making
    of the She-Male* (Raymond),
        234-239
Transsexuality, 37n13
    as diagnostic category, 32
Transsexuals
    Butler's theory of gender
        performativity and, 32-33
    lesbian communities and, 19-22
    male-to-female, 5
Transvestitism, 29-30
Traub, Valerie, 112
The Truth About Jane, 118-120
2Grown Project, 47-49
Two-spirit women, 127-128. *See also*
        Lesbians
    as collective identity, 133-135
    coming out *vs.* becoming, 135-136

    confronting compulsory
        homonormativity and,
        136-138
    discrimination and, 142-146
    effect of culture for, 138-140
    family and community for, 140-142
    indigenous resistance and, 130-133
    serving community and, 129-130
    spirituality and, 128-129

An Unexpected Love, 120-121

Vibes. *See* Gaydar; Gay vibes

The Well of Loneliness, 94,96,97
What Makes a Family, 120-121
White, Hayden, 93
Wilson, Alex, 128
Wittig, Monique, 111
Wolf, Deborah Goleman, 90
Womyn-born womyn criterion,
        9-10,182,183-184
Woolf, Virginia, 190-193
Wounded attachment, 96-97

Youth
    intersex, 45
    transgender-identified, 47-49
        Reid, 56-64
        Samantha, 49-56

Zita, Jacquelyn N., 31

BOOK ORDER FORM!

Order a copy of this book with this form or online at:
http://www.HaworthPress.com/store/product.asp?sku= 5799

Challenging Lesbian Norms
Intersex, Transgender, Intersectional, and Queer Perspectives

____ in softbound at $19.95 ISBN-13: 978-1-56023-645-0 / ISBN-10: 1-56023-645-0.
____ in hardbound at $39.95 ISBN-13: 978-1-56023-644-3 / ISBN-10: 1-56023-644-2.

COST OF BOOKS _____

POSTAGE & HANDLING _____
US: $4.00 for first book & $1.50
for each additional book
Outside US: $5.00 for first book
& $2.00 for each additional book.

SUBTOTAL _____

In Canada: add 7% GST. _____

STATE TAX _____
CA, IL, IN, MN, NJ, NY, OH, PA & SD residents
please add appropriate local sales tax.

FINAL TOTAL _____
If paying in Canadian funds, convert
using the current exchange rate,
UNESCO coupons welcome.

❑ BILL ME LATER:
Bill-me option is good on US/Canada/
Mexico orders only; not good to jobbers,
wholesalers, or subscription agencies.

❑ Signature _____

❑ Payment Enclosed: $_____

❑ PLEASE CHARGE TO MY CREDIT CARD:
❑ Visa ❑ MasterCard ❑ AmEx ❑ Discover
❑ Diner's Club ❑ Eurocard ❑ JCB

Account #_____

Exp Date_____

Signature_____
(Prices in US dollars and subject to change without notice.)

PLEASE PRINT ALL INFORMATION OR ATTACH YOUR BUSINESS CARD
Name
Address
City   State/Province   Zip/Postal Code
Country
Tel   Fax
E-Mail

May we use your e-mail address for confirmations and other types of information? ❑Yes ❑No We appreciate receiving your e-mail address. Haworth would like to e-mail special discount offers to you, as a preferred customer.
We will never share, rent, or exchange your e-mail address. We regard such actions as an invasion of your privacy.

Order from your **local bookstore** or directly from
The Haworth Press, Inc. 10 Alice Street, Binghamton, New York 13904-1580 • USA
Call our toll-free number (1-800-429-6784) / Outside US/Canada: (607) 722-5857
Fax: 1-800-895-0582 / Outside US/Canada: (607) 771-0012
E-mail your order to us: orders@HaworthPress.com

For orders outside US and Canada, you may wish to order through your local
sales representative, distributor, or bookseller.
For information, see http://HaworthPress.com/distributors

(Discounts are available for individual orders in US and Canada only, not booksellers/distributors.)

The Haworth Press Inc.

Please photocopy this form for your personal use.
www.HaworthPress.com

BOF06